The Cognitive Partnership

AI and the Future of Intelligent Work

Carl Andersson

Nutropic AI · 2026

Contents

Copyright 1
Preface 3

I The Framework 5

1 The Problem With How Most Organisations Think About AI 7
2 The Four AI Operating Models 17
3 Paired Cognition: The Most Powerful Way to Use AI 27
4 Delegation, Agency and Embedding: When AI Works Without a Partner 35
5 Choosing the Right Operating Model 43

II Inside Paired Cognition 53

6 Externalised Cognition: Thinking Out Loud With AI 55
7 Distributed Cognition: The Three-Node Model of Human-AI Teams 65
8 Building Organisational Memory: The Knowledge Base 75
9 How AI Joins Your Conversations: Direct, Observed and Assisted 85
10 The Seven Phases of a Paired Cognition Workflow 95
11 Directed Execution vs. Task Delegation: Why Context Changes Everything 103
12 The Reflection Phase: Why the Work Is Not Done When the Output Is 111

III Making It Happen 121

13 AI Strategy: From Readiness Assessment to Actionable Roadmap 123
14 Workflow Mapping: The Foundation of Intelligent AI Integration 135
15 Building AI Automations That Connect to Your Business 145
16 Training Teams for Real AI Adoption 155
17 Embedding AI in Software Systems as Core Capability 165
18 Piloting AI: How to Prove Value Before You Scale 175

IV Across Industries **185**

19 AI as a Cognitive Partner in Finance 187
20 AI-Augmented Underwriting and Claims in Insurance 197
21 AI in Legal Practice: Research, Contracts and Strategy 205
22 Clinical AI: Decision Support That Keeps Humans in Charge 217
23 AI in Marketing and Communications: Two Registers, One Discipline 227
24 AI in Technology Teams: Architecture, Product and Operations 239
25 AI in Education: Teaching, Curriculum and Administration 249

V The Organisation and Its People **263**

26 Orchestration: Composing AI Methods Across an Organisation 265
27 Governance: Oversight, Accountability and Control 275
28 Why AI Adoption Fails, and How to Make It Stick 285
29 The Human Side of AI Partnership: What Changes When Teams Work With AI 295
30 The Cognitive Future: Where Paired Cognition Goes Next 305
31 What to Do Now: Immediate Actions for Organisations 315

About the Author 325
Acknowledgements 327
References 329

Copyright

ISBN 978-1-7646736-0-0 (print)
ISBN 978-1-7646736-1-7 (ebook)
ISBN 978-1-7646736-2-4 (audiobook)

First edition.
Nutropic AI, Melbourne, Australia, 2026.

This book was produced in collaboration with AI systems. The framework, research, argument, direction and editorial judgement are the work of the author. AI systems were used to draft and edit the manuscript under close author direction throughout the writing process.

Preface

The idea arrived as a way of working before it arrived as anything resembling a theory, and most of what follows has the texture of that order.

The project was an orchestration layer for software development lifecycle management. What shifted early was not the output but the method: less time typing, more time in conversation with my AI counterpart. Challenges started as open questions, exploratory exchanges that found their own direction. The conversations became requirements, then specifications, then design decisions. The progression felt natural because it is natural. Human capacity for verbal discourse substantially exceeds our ability to express things in writing. AI on the other hand has an unbelievable capacity to elicit reasoning and surface information signals, even from a brain dump full of typos.

The working relationship became genuinely symbiotic, collecting and organising information, making and documenting key decisions, then executing against them. That efficiency was not specific to the software project. Within weeks that same method had extended naturally into all my other work. Today practically all my communication with AI is verbal or using voice-to-text.

The broader conversation about AI has not caught up. Its dominant frame is automation and autonomy: doing the same work cheaper, faster and at greater scale. The efficiency gains are real and not to be dismissed. But the fixation on doing the same thing more efficiently consistently crowds out a more consequential question: how AI can make human reasoning better, not just faster. The most powerful thing AI can do is improve the quality of human decisions. Automation produces efficiency. Paired Cognition, a sustained partnership between a person and an AI engaged with the same problem throughout, produces capability. These are not the same gain.

It matters because the quality of decisions is where consequential work lives, and the organisations I work with are not short of execution capacity. Many are short of the kind of thinking that produces the right question before the wrong answer gets built. If AI can genuinely raise the ceiling on human reasoning, that changes what is possible in the work that matters most. An AI strategy built around automation and delegation makes that gain invisible.

This book is for the person who already suspects the automation frame is incomplete. Not the person who needs to be convinced that AI is significant; that argument has been settled. The person sitting with a complex problem, trying to think it through rather than execute it away, who wants to know how AI can genuinely help with that kind of work. That is the reader this book was written for.

A note on what the book is not. The tooling available when these pages were written will have changed by the time you read them; a book organised around specific platforms or models would be obsolete before it reached printing. What changes more slowly is the question of how humans and AI should relate to each other when the work is complex, consequential and resistant to full specification in advance. That is the question this book addresses.

This book was written through the method it describes. The framework, arguments, structure and every editorial decision are my own. The execution of that thinking into prose was accomplished through Paired Cognition, with AI as a writing partner. The book's argument is that the distinction between human-authored thinking and AI-executed writing is not a contradiction to be explained away. It is a description of how the model works at its best.

Carl Andersson
Melbourne, 2026

Part I

The Framework

The four AI operating models, what distinguishes them, and how to choose between them workflow by workflow.

Chapter 1

The Problem With How Most Organisations Think About AI

> *The more efficient you are at doing the wrong thing, the wronger you become. It is much better to do the right thing wronger than the wrong thing righter.*
>
> — Russell L. Ackoff

Most organisations are spending more on AI than they have ever spent on any new technology, and getting back less than they expected. The capability is real. The budgets are real. The executive attention is real. And yet the gap between what AI can do in a demonstration and what it actually delivers inside a working business has become one of the defining frustrations of the present time.

The scale of that gap is now well documented. MIT's *GenAI Divide* study (MIT NANDA Initiative 2025), which combined an analysis of three hundred enterprise deployments with leader interviews and a workforce survey, found that roughly 95 per cent of organisations had seen no measurable profit-and-loss impact from their generative AI pilots, despite combined enterprise spending in the order of $30 to $40 billion. The capability had arrived. The returns, for almost everyone, had not.

The problem is not the technology itself but the thinking around it.

This is a publication about that thinking. Specifically, it is about a class of structural decisions that most organisations never make explicitly, decisions that quietly determine whether AI adoption produces transformative outcomes or expensive disappointment. Although these decisions look technical at first, they are operational at their core, and they begin with a question almost nobody asks clearly: when AI is involved in the work, who is actually doing the thinking, and how is responsibility shared between humans and the system?

That question sounds abstract, but the answer to it determines what tools are bought, how teams are trained, what governance is needed, what risk profile the organisation takes on and ultimately what kind of value AI is allowed to produce.

Organisations that answer it deliberately, workflow by workflow, end up with AI that compounds. Organisations that never answer it end up with AI that disappoints, and with no clear account of why.

1.1 A monolith is not a strategy

The most common and most consequential mistake is to talk about AI as if it were a single technology playing a single role. Leaders speak about "rolling out AI" or "AI in our operations" as if the choice were binary, as if the only decision were whether to adopt or not. The framing fits neatly on a slide, and that is part of why it obscures something fundamental about how AI actually shows up in a real organisation.

AI is not one thing inside an organisation. The way it participates in a strategic conversation between two senior professionals is categorically different from the way it processes documents asynchronously in a queue overnight. The way it sits inside a customer-facing application is different again. These are different activities, not faster and slower versions of the same one: their inputs, their outputs, their success criteria and their failure modes are all different.

Treating them as the same leads to predictable damage.

> *A wealth advisory practice buys a productivity assistant marketed for general professional work and quietly expects it to handle compliance review. A claims operation invests in a fully automated triage agent and is surprised when senior adjusters refuse to trust its outputs on bodily injury matters. An engineering team rolls out a code-generation tool firm-wide and cannot understand why the architects who do the highest-leverage work seem least enthusiastic.*
>
> *None of these are tooling failures. They are category errors about what role AI is actually playing in the work.*

Until an organisation can name which kind of relationship it is designing in a given workflow, it cannot make sound choices about anything downstream: not tools, not training, not governance, not measurement. The monolith framing forces every subsequent decision onto a foundation that does not exist.

1.2 When automation becomes the only lens, quality disappears from view

Once AI has been collapsed into a single concept, the question of value collapses with it. Most organisations default to a narrow framing. AI is for cost reduction and speed. The implicit model is automation. Replace humans with AI. Do more with less. Finish faster. Spend less on labour.

That framing is not wrong in every case, but it is dangerously incomplete in the cases that matter most.

When automation becomes the default lens, entire categories of value disappear from view. The most significant of these is decision quality. There are workflows, often the most economically important an organisation has, where the goal is not to do the same work cheaper but to do better work than either a human or an AI could produce alone. In these workflows, the combination of human judgement and AI capability produces decisions demonstrably superior to anything either party reaches independently. Quality is the point, and speed comes along as a side effect of pursuing it well.

This is a different proposition from automation, and it requires different language to describe. It is not about output per hour. It is about the depth and defensibility of a complex judgement made under real conditions, with real consequences, by a person whose name is on the file. AI's contribution in this kind of work is to extend the practitioner's reasoning, not to replace it: to hold more context, surface more options, test more assumptions and maintain the through-line of an analysis the practitioner could not hold alone in the same time.

Organisations fixated on automation never look for these outcomes, so they never find them, and they conclude that AI is overhyped. The conclusion is wrong. The experience is real. They have built a value lens that cannot see the value most worth pursuing.

1.3 The bottleneck has shifted from capability to integration

There is now a striking and persistent gap between what frontier AI systems can do in benchmark conditions and what organisations actually achieve when they deploy these systems in real workflows. Headline capabilities advance every few months. Adoption quality advances much more slowly. The technology is doing roughly what it claims. The gap exists because capability and adoption quality are governed by entirely different things.

Capability is determined by the model and its tools. Adoption quality is determined by how the model is integrated into the work humans actually do. A highly capable system embedded in a poorly designed workflow performs worse than a less capable system embedded in a well-designed one. Five years ago, the limiting factor was what AI could do. Today, in most professional settings, the limiting factor is how the work around AI is structured.

This is uncomfortable for organisations that have approached AI as a procurement exercise. Buying access to a capable model does not produce value. Designing workflows that use that capability well does. Most procurement processes are not equipped to evaluate workflow design, and most vendors do not sell it. The capability-adoption gap persists because almost nobody is paid to close it.

> *Two specialty insurers buy access to the same frontier model on the same day. Six months later, one has measurably tighter underwriting on its largest risks, with senior underwriters writing better-reasoned referrals in less time. The other has a sprawling collection of disconnected experiments and a growing sense that the model is "not quite there yet."*
>
> *Both have the same capability. Only one has done the work of designing how that capability fits into the way decisions are actually made.*

The broader data tells the same story. McKinsey's *State of AI* report (McKinsey & Company 2025c) found that 88 per cent of organisations now use AI in at least one business function, yet only 39 per cent attribute any EBIT impact to it, and just 6 per cent of so-called "AI high performers" attribute more than 5 per cent of EBIT to it. Of the twenty-five organisational attributes McKinsey tested, the single largest determinant of whether an organisation realised EBIT impact was fundamental workflow redesign. Access to the model is now nearly universal. Redesigning the work around it is not.

1.4 What fills the silence?

The third error is what many organisations do in response to the first two. Recognising, often correctly, that something more is needed than tool procurement, they launch transformation programmes. The programmes produce vision statements, capability maps, centres of excellence and pilot portfolios. They make impressive presentations. They generate organisational motion.

They often fail to change how work is actually done.

The pattern is consistent enough to deserve a name. Transformation theatre is the production of visible activity around AI without the underlying operational design that would make AI valuable. Workshops happen. Frameworks are introduced. Executives become fluent in the vocabulary. Strategy decks acquire a dedicated AI section. Meanwhile, in the actual workflows where decisions are made and outputs are produced, the relationship between humans and AI remains undefined.

> *A regional bank announces an AI strategy, stands up an AI council, runs a portfolio of pilots across functions and reports quarterly to the board on capability uplift.*
>
> *Two years in, every credit memo is still drafted the same way it was before. The pilots that worked never made it into the way the work is done. The pilots that did not work are quietly forgotten. The vocabulary of transformation has spread through the organisation without the operational design to make it real.*

The missing element in scenarios like this is not vision. It is not governance. It is not talent. Those things matter, and they are not the bottleneck. The missing element is operational design at the level of specific workflows: a clear, deliberate decision about how humans and AI will share the work in each context where AI is used. Without that decision, every other investment is built on sand. Strategy without operating model selection is theatre. Governance without operating model selection is bureaucracy. Training without operating model selection is interface tuition.

1.5 The missing primitive is the operating model

This publication introduces a concept designed to fill the gap. It is called an AI operating model. The phrase is deliberately chosen, and it is worth pausing on.

It is not an "AI model" in the machine learning sense. It is an operating model: a structural definition of the relationship between humans, AI and the work being done. The decisive axis is decision-making responsibility. Who decides? Who executes? Where does human judgement enter the work, and where does it not? Those questions sound philosophical until they are asked of a specific workflow. Asked there, they cut through nearly every confusion that surrounds AI adoption.

Four operating models cover the practical landscape.

Paired Cognition is a sustained partnership in which a human practitioner and AI maintain a shared mental model throughout the work, with joint engagement on decisions and AI-led execution validated by the human. It is the model most under-recognised in current practice and the one most relevant to complex professional judgement.

Task Delegation is a workflow in which a human decomposes the work into discrete tasks, assigns those tasks to AI for execution against a brief and reviews the output. It is the model most organisations are using when they think they are doing AI adoption.

Autonomous Agency is a workflow in which AI operates independently within objectives and guardrails set by humans. The human is supervisor rather than collaborator. This is the model behind the most advanced agentic systems now reaching production.

Embedded Intelligence is a workflow in which AI functions as a component inside automated infrastructure: configured at design time, invoked without a human interface during operation and governed entirely through the surrounding software architecture.

These four models are not mutually exclusive, and they are not a maturity ladder. A single organisation, and often a single team, will use different operating models for different workflows.

A mid-sized law firm might use Paired Cognition for litigation strategy on a contested matter, Task Delegation for first-pass drafting of routine engagement letters, Autonomous Agency for ongoing monitoring of clauses across a contract portfolio and Embedded Intelligence for the document classification that runs silently inside the firm's matter management system. The four models coexist. Each one fits a particular kind of work. None of them substitute for the others.

The operating model is chosen per workflow, deliberately, based on what the work actually requires. That sentence, read carefully, contains most of the discipline this publication is trying to teach.

1.6 What the framework changes

Once an organisation can name its operating models, almost every other AI decision becomes clearer.

Tool selection follows from the operating model rather than driving it. Training requirements differ sharply across models: Paired Cognition demands skill in externalising thought and partnering with a system that needs context; Task Delegation demands skill in decomposing work and writing briefs that a competent stranger could execute. Governance differs. Risk profiles differ. Measurement differs. The operating model is not a layer added on top of these decisions but the foundation that makes them coherent in the first place.

The framework also resolves the automation fixation by making different kinds of value visible. Where Task Delegation maps closely to the classical productivity story of cost and speed, Paired Cognition is concerned with decision quality, defensibility and the capacity to do work that neither party could produce alone. Autonomous Agency unlocks scale in narrow, well-bounded domains. Embedded Intelligence is essentially software architecture with AI components replacing conventional logic, evaluated on the same terms as any other production system.

Naming the four models lets organisations set realistic expectations and pursue the right outcomes for each. It also makes it possible, finally, to have a coherent conversation about where the value is actually coming from.

1.7 A note on how this publication is organised

The publication runs to 31 chapters in five parts.

Part I establishes the framework, introducing all four operating models and the axis of decision-making responsibility that differentiates them. It closes with the discipline of choosing the right model for a given workflow.

Part II goes deep on Paired Cognition: the cognitive architecture that underpins it, the role of organisational memory, the conversation types that feed it and the seven-phase workflow that structures it in practice. It is the longest part of the book because it is where the most under-recognised value sits.

Part III turns framework into adoption: strategy, workflow mapping, automation, training, embedding AI in software and the design of pilots that prove value rather than perform it.

Part IV applies the framework across seven professional industries: finance, insurance, legal, clinical practice, marketing and communications, technology and education. Readers from any of these will find their context here.

Part V addresses the organisational level: orchestration, governance, why adoption fails, the human dimension of working with AI at scale, the cognitive future and a practical action agenda for the next 90 days.

A reader can enter at any point. The chapters build on each other, and the practical recommendations later in the book assume the conceptual foundation laid in Part I.

1.8 Who this is for

This publication is written for the people actually making AI adoption decisions inside real organisations. That includes leaders responsible for setting direction, practitioners responsible for designing the workflows where AI lives and teams responsible for doing the work AI is meant to support.

It is not written primarily for technologists, although technologists will find it useful. It is written in the language of work, decisions and outcomes: the terms in which AI either succeeds or fails inside an organisation.

The argument is that AI adoption is an operating model problem before it is a technology problem, and that organisations which make their operating models explicit will pull steadily ahead of those that do not. Not because they have better models or larger budgets, but because they have stopped trying to answer questions about tools, training and governance without first answering the question those decisions depend on.

The work begins with a single question, asked clearly for each workflow that matters: what is the right relationship between humans and AI here, and how will it be designed for? Everything in the chapters that follow is in service of answering that question well, and the answer is rarely the one already on the slide.

CHAPTER 2

The Four AI Operating Models

> *Don't say: 'They must have something in common, or they would not be called "games"' — but look and see whether there is anything common to all. For if you look at them, you won't see something that is common to all, but similarities, affinities.*
>
> — Ludwig Wittgenstein

A finance team rolling out a copilot for analysts, a legal department triaging contracts overnight and an insurer embedding a classifier inside its claims pipeline tend to appear on the same slide. They are funded from adjacent lines in the same budget. They are sometimes governed under the same policy. They are spoken about in the same breath.

They do not belong to the same category of thing.

The strategist working through a market entry decision with an AI partner is doing different work from the operations team running a nightly classification job. The relationship between human, machine and outcome is structurally different in each case. The previous chapter named this confusion. This chapter takes it apart.

The vocabulary begins with one recognition: AI inside an organisation does not have a single operating mode but four, and they are not interchangeable.

2.1 What an operating model actually describes, and what it does not

An AI operating model describes how AI operates within a workflow. The phrase is deliberately distinct from the machine learning sense of "model," which refers to a trained system such as a transformer or a classifier. Here the term carries its meaning from business architecture. It names the relationship between humans, AI and the work being performed.

This is a structural question, not a technical one. It does not ask which vendor was selected, which model is hosted or which interface was rolled out. It asks how the work is shaped: where decisions are made, who makes them and how the AI participates in the activity that produces the outcome.

One variable matters more than the others. The single load-bearing dimension is decision-making responsibility. Every other dimension follows from that root question: who executes, how often a human is involved, what oversight applies, where governance attaches. Locate decision-making responsibility, and the rest of the operating model falls into place.

At one end of the axis, humans and AI share decision-making responsibility throughout the work, functioning as cognitive partners with both engaged from beginning to end. At the other end, AI carries no decision-making weight visible to a human at runtime, because the decisions were made once at design time and the system now executes against them without ongoing human involvement.

Between those poles sit two intermediate configurations. Each is useful in its own contexts. None of the four is a stage on the way to the others.

2.2 A single spectrum, four shapes

The four operating models are Paired Cognition, Task Delegation, Autonomous Agency and Embedded Intelligence. They form a spectrum, ordered by how decision-making responsibility shifts from being shared with a human to being absorbed into automated infrastructure.

Operating Model	Decision-Making	Execution	Human Involvement
Paired Cognition	Joint shared responsibility throughout	AI-led with human validation	Continuous partnership
Task Delegation	Human decomposes and assigns	AI executes discrete tasks	Assignment and review
Autonomous Agency	AI independent decisions within objectives	AI alone	Objective-setting and oversight

Operating Model	Decision-Making	Execution	Human Involvement
Embedded Intelligence	System (AI as component)	Automated	Design-time only

Reading the table from top to bottom traces a quiet handover. Decision-making moves from a shared activity to an assigned one, then to a delegated one and finally to a designed-in property of the system. Human involvement moves from continuous to occasional to architectural. Each row is a coherent way of working in its own right. None of them is a halfway house to a different row.

The order matters less than the distance between rows. Paired Cognition and Embedded Intelligence are not competing approaches to the same problem. They answer different problems. The reason to put them on a shared axis is to make the shape of the difference visible, not to suggest that organisations should pick a position on the line and stay there.

2.3 Paired Cognition is a sustained partnership, not a faster version of the same work

Paired Cognition is a sustained cognitive partnership in which a human and an AI maintain a shared mental model throughout the work. Neither party disengages. The human contributes judgement, domain expertise and contextual reasoning. The AI contributes the ability to process information broadly, hold consistency across long workstreams and generate alternatives at scale. The two think together.

The defining property of this model is not speed. It is quality. A senior strategist working with an AI partner on a market entry analysis is not producing the same memo faster. She is considering more options. She is stress-testing assumptions more rigorously. She is arriving at recommendations that draw on both human and machine intelligence and would not have been reachable by either party alone.

> *The senior strategist arrives at a Tuesday morning session with a half-formed view about a regional expansion. Over the next two hours she externalises that view and tests it against everything the AI can hold in mind at once: the compet-*

itive landscape, the regulatory environment in three jurisdictions, the historical pattern of similar entries by similar firms, the operational implications under three different demand scenarios.

The recommendation that lands on the partner's desk on Thursday is not a faster version of what she would have written alone. It is a different recommendation, drawn from a wider view.

This is the model that organisations most often misread. It is mistaken for a more elaborate version of Task Delegation, in which the prompts are longer and the model is better, when in fact the relationship differs in kind. Both the human and the AI stay in the work, so the output is the trace of a partnership rather than the result of an instruction.

Paired Cognition fits situations where decision quality is the actual ambition: complex strategic problems, high-stakes legal reasoning, clinical decision support in ambiguous cases, the design of novel financial products, the framing of a difficult negotiation. It is the most under-recognised of the four models in current practice, and the model the rest of Part II is dedicated to. The pattern is visible in the usage data as well as in the work itself. Anthropic's *Economic Index* (Anthropic 2025a) reported that on its consumer service, roughly 52 per cent of conversations followed an augmentation pattern (iterative thinking with a human in the loop) against 45 per cent of an automation pattern. On programmatic API traffic the ratio inverted. Where humans choose how to engage, partnership tends to win; where AI is wired into pipelines, automation does. Both findings are consistent with the four-model spectrum, and both suggest that Paired Cognition is doing more of the cognitive work in everyday use than the prevailing narrative admits.

2.4 Task Delegation is delegation in the ordinary sense of the word

Task Delegation describes a different shape of work. The human decomposes a piece of work into discrete tasks, then assigns each task to an AI for execution. Because the AI executes without broader context of the overall project, decision-making responsibility stays with the human while the AI carries only the responsibility for execution.

The relationship resembles that of a senior professional with a capable but uncontextualised assistant. The assistant can perform the task well when the task is well-specified. The assistant cannot be expected to understand the case the task sits inside, the strategy the case forms part of or the client relationship the strategy is shaped by.

Concrete examples are easy to find. A lawyer drafting a settlement agreement might hand off the redlining of a specific clause against a precedent, the formatting of an exhibit list or the summarisation of a single deposition. A finance analyst might delegate the construction of a particular chart from a defined dataset. A communications team might hand off the production of channel-specific variants of an already-agreed message. Each task is well-bounded. Each has clear success criteria. Each can be performed without the broader picture.

The trade-off is that the human shoulders the cognitive load. They must know the work well enough to break it into the right pieces. They must judge which pieces suit delegation and which do not. They must reassemble the outputs into a coherent whole and take responsibility for the result. Task Delegation is efficient for execution. It does not augment thinking the way Paired Cognition does, because the AI is never inside the thinking.

This is the model most teams already know, because it maps most neatly onto existing habits of work. That familiarity is also its trap. Task Delegation becoming the unexamined default, the answer to every question of the form "where could AI help here?", is one of the recurring patterns this book will return to.

2.5 Autonomous Agency moves the human from collaborator to supervisor

Autonomous Agency moves further along the spectrum. The AI operates independently against defined objectives, with minimal human involvement during execution. Humans set the objectives, the constraints and the guardrails. The agent then works inside those boundaries and makes its own decisions about how to proceed. Decision-making responsibility, within the agreed scope, sits with the AI.

A customer service agent in a mid-sized telecommunications business handles a defined class of routine queries end to end.

A customer messages about a billing query. The agent verifies the account, retrieves the recent invoices, identifies the source of confusion, drafts a response, applies a credit if the situation falls within pre-agreed limits and closes the case. A human supervisor sees the case only if the agent's confidence falls below a threshold, the customer escalates or a sample is pulled for quality review. Most cases never touch a human.

The same pattern shows up in operations, where agents triage incoming alerts, gather context from the relevant systems, apply known remediations and escalate the genuinely novel cases to a human. It shows up in finance operations, where agents reconcile transactions, investigate exceptions and clear or refer them according to scope. It shows up in document-processing pipelines, where agents read incoming documents, classify them, extract the relevant fields and route the result.

The human's role shifts. From collaborator inside the work, the human becomes a supervisor sitting outside it. Governance becomes the work of setting the right objectives, defining the constraints carefully and reviewing outputs at the boundaries rather than within the moment-to-moment activity. This is a substantial change in how a manager spends time. It also changes the failure modes. Errors in objective specification or guardrail design have larger consequences than errors in any single task, because they are not corrected mid-stream.

Autonomous Agency is the model most often confused with the others, because the language of "AI agents" travels widely and is rarely tied to a specific operating model. The adoption picture is uneven but no longer marginal: McKinsey's late-2025 *State of AI* survey (McKinsey & Company 2025d) reported that 23 per cent of organisations were scaling AI agents in at least one function and a further 39 per cent were experimenting, with the highest concentrations in IT service desks, internal knowledge retrieval, engineering copilots and customer operations. The shape of that distribution matters. Agents are landing first in places where the work is well-bounded, the objectives are stable and the cost of a misjudged action is contained, which is exactly where the operating model fits. A later chapter develops it in depth, including the question of which kinds of work belong to it and which do not.

2.6 Embedded Intelligence is AI as part of the machinery, not as a participant in the work

At the far end of the spectrum, Embedded Intelligence places AI inside automated infrastructure as a component. The system is invoked by other automated processes. There is no human interface during operation. The human's role is limited to designing and configuring the system, after which it runs as part of the underlying machinery of the organisation.

A claims processor that uses a model to classify documents as they arrive sits in this category. So does a fraud detection layer that scores transactions in real time, an underwriting service that pre-scores submissions before they reach a human or a background service in a publishing platform that performs routine authoring work asynchronously in response to queued tasks. Each is conceptually similar to traditional software architecture, with AI components replacing conventional logic at specific points.

Because decision-making was settled at design time, operational decisions function as policies rather than as judgements made in the moment. Governance happens through software engineering disciplines: testing, monitoring, versioning, change control, incident response. The model behaves more like a feature of the platform than a participant in the work, and the people responsible for it are typically engineers and product owners rather than the professionals whose work it supports.

This is the model that disappears from view most easily. When it is working, it is invisible. That invisibility is part of its value. It is also why Embedded Intelligence is sometimes treated as the only "real" form of AI inside an organisation, the only one that scales without people in the loop. That framing misses the point. Each of the four models scales in its own way. Embedded Intelligence scales by removing the human from the runtime. Paired Cognition scales by lifting the quality of decisions a human can make. Both are real, and neither is the default.

2.7 Not a ranking, a portfolio

It is tempting to read the four models as a maturity ladder, with Paired Cognition as the sophisticated end state and Embedded Intelligence as a humbler precursor.

The opposite reading is equally tempting: Autonomous Agency as the pinnacle of automation, with the other three as steps along the way to fully autonomous work. Both readings are wrong.

The models are not ranked by quality. Each is the right choice in specific contexts, and applying the wrong one is harmful in either direction. Forcing Embedded Intelligence onto a workflow that needs Paired Cognition produces brittle systems and hollow decisions. Forcing Paired Cognition onto a workflow that should be embedded produces expensive, slow processes and frustrated professionals.

The discipline is matching the model to the work. That depends on understanding the work first and the technology second.

A well-run organisation will operate several of the four in parallel, applied to different workflows according to fit. A bank might use Paired Cognition for credit committee analysis, Task Delegation for analyst-driven research, Autonomous Agency for certain treasury operations within strict limits and Embedded Intelligence for transaction monitoring. Each is the right choice in its own context. None of them displaces the others.

The law firm example from the previous chapter shows this coexistence directly; subsequent chapters develop each model in turn.

Coherent adoption looks like a portfolio of operating choices rather than a single AI strategy applied uniformly: each model named, each governed in the way it requires, each chosen because it fits the work it is doing.

2.8 Paired Cognition is where the book opens its engine

Of the four, one model carries the heaviest argument and most of the latent value, and one model is also the most consistently misread. Both are Paired Cognition. The next chapter takes it apart in detail: what a shared mental model actually contains, how it is built between human and AI, why it produces qualitatively different output rather than merely faster output and where it fits in the work that pays the bills inside professional firms.

The other three models receive the same treatment in Chapter 4. Chapter 5 then turns to the discipline that ties Part I together: choosing the right model for the work in front of you, the criteria that make the choice deliberate and the symptoms that signal a mismatch.

Everything that follows assumes that an AI deployment can be named, that the question "which operating model is this?" has an answer and that the answer matters. Most current adoption programmes cannot answer it. The work of this book is to make the answer routine, and the route to that begins with the model the dominant narrative has tried hardest to ignore.

Chapter 3

Paired Cognition: The Most Powerful Way to Use AI

> *Weak human + machine + better process was superior to a strong computer alone and, more remarkably, superior to a strong human + machine + inferior process.*
>
> — Garry Kasparov

There is a particular kind of professional moment that does not yield to faster typing. A senior practitioner is holding several incompatible considerations in working memory at once. The matter in front of her is not novel in the abstract, but it is novel in this configuration: these constraints, this client, this regulatory posture, this prior decision the firm made 18 months ago and would prefer not to revisit. The output that finally lands on the page will look, to an outside reader, like a crisp piece of work. What it actually represents is a long arc of judgement, with a dozen options entertained and discarded along the way.

This is the work that pays the bills in most professional firms. It is also the work the dominant narrative around AI has the most trouble describing. Productivity assistants speed up the typing. Automation handles the routine. Neither framing has anything useful to say about what is happening in that practitioner's head as she works the problem. Neither, applied as a default, gets within reach of the value AI can actually contribute to that kind of thinking.

Paired Cognition is the operating model for that kind of work. It is the most under-recognised of the four, and where it fits, the most consequential.

3.1 The point is decision quality, not speed

The first thing to be clear about is what Paired Cognition is for. The default lens organisations bring to AI is some combination of cost and speed: the same work, cheaper or faster. Paired Cognition does often produce both. But framing it that way mistakes a side effect for the point.

Paired Cognition is about decision quality.

In the workflows where it fits, the value is not that the practitioner reaches the same answer in less time. The value is that she reaches a better answer than she would have reached alone: better grounded, more carefully considered, tested against more alternatives, defensible in more directions. The combination of human judgement and AI capability produces outcomes that neither party could reach independently. Speed and cost gains follow, but they are downstream of a more fundamental shift in how the thinking itself gets done.

This reframing changes what an organisation looks for when it asks whether AI is working. An organisation measuring only output per hour will not see the value of Paired Cognition even when it is happening. The practitioner is producing fewer documents; she is producing better ones. The throughput metric registers a small gain. The defensibility, depth and downstream economic consequence of the decisions inside those documents register no signal at all. The organisation concludes the AI is "fine, not transformative," and quietly looks elsewhere.

What was actually happening was the highest-leverage form of AI use the organisation had access to, measured against the wrong thing.

3.2 A shared mental model is the defining feature

The mechanic that distinguishes Paired Cognition from the other three operating models is the shared mental model that human and AI maintain throughout the work.

That phrase needs unpacking, because it carries a lot of load.

A shared mental model contains the why of the work, not only the what. It captures the goal, the constraints, the audience, the strategic intent, the prior decisions that govern this matter, the standards the output has to meet and the things already ruled out. It captures the practitioner's read of the situation: which considerations are live, which are settled, which are politically sensitive, which are technically subtle. It is built deliberately during the early phases of the engagement and referenced continuously thereafter.

Building it requires investment from both sides. The human contributes domain expertise, contextual sensitivity, judgement about edge cases, awareness of organisational and political dynamics and the lateral thinking that connects signals which are not formally connected. The AI contributes structuring capacity, the reference breadth to surface relevant precedent at scale and the discipline of holding consistency across long bodies of reasoning. Each contribution has a place, and neither substitutes for the other.

Once established, the shared model behaves like a working memory across the partnership. The practitioner can step back into a thread the next morning and find the context preserved. The AI can generate options that respect constraints established hours earlier in the conversation, without being re-briefed each time. The model becomes something like a third presence in the room, holding the accumulated reasoning of the engagement in a form neither party alone would carry forward reliably.

This is what distinguishes partnership from delegation. In delegation, the human briefs the AI, receives an output and judges it against the brief. In partnership, the human and the AI think together, with the model of the work itself shared between them. The output is not a deliverable handed across a boundary. It is the product of a continuous conversation in which both participants have stake.

3.3 Augmented intelligence is the goal Paired Cognition implements

Paired Cognition is best understood as a practical implementation of augmented intelligence. The phrase predates the current AI moment; it was in use in research circles long before large language models. It has acquired sharper meaning now that the underlying capability is real.

Augmented intelligence describes a system in which human cognition is extended, not replaced, by machine capability. The combination is designed to outperform either party operating alone. It is not the same as automation, which removes the human from the loop. It is not the same as assistance, which leaves the human's reasoning unchanged and only speeds up the surface execution. It is a third thing, in which the structure of the thinking itself is different because both parties are participating in it.

Human judgement remains central to that structure. Domain expertise, ethical reasoning, contextual sensitivity and the trained instinct that recognises when something feels wrong are not capabilities current AI systems possess in full. They are also, more importantly, not capabilities organisations should be willing to outsource even where AI can approximate them. Professional accountability does not transfer to a model. The practitioner whose name is on the file is still the practitioner whose name is on the file.

The AI side of the partnership brings complementary strengths. It processes information broadly, recalls precedent without fatigue, generates alternatives at scale and applies consistent structure across long documents. These capabilities, applied in isolation, produce competent but generic work. Applied within a shared mental model, they produce work that feels genuinely informed: work that knows what it is for.

The combination differs in kind from either input. A senior practitioner working alone produces good work, slowly. A generic AI working alone produces fast work that misses the point. The pair, properly structured, produces work that is both fast and deeply contextual. The whole exceeds the sum of its parts in a way pure automation never achieves, because the mechanism is not addition. It is the formation of a working unit that thinks differently from either component.

The empirical signal here is sharper than the rhetoric usually allows. In a pre-registered field experiment with 758 BCG consultants, Dell'Acqua and colleagues (2025) found that those using GPT-4 inside the AI's capability frontier completed 12 per cent more tasks, worked 25 per cent faster and produced output rated 40 per cent higher in quality than the control group. Outside that frontier, the same AI access made consultants 19 percentage points *less* likely to reach the right answer. The lesson is not that AI is universally helpful or universally hazardous; it is that the gains live inside a specific envelope, and a partnership that knows the shape of that envelope is the difference between augmentation and harm.

3.4 Iron Man is not Tony Stark, and Iron Man is not Jarvis

The clearest popular image of this dynamic comes from a comic-book franchise. Tony Stark, in the Marvel films, is a brilliant engineer. Jarvis, his AI system, is an extraordinary assistant. While each is impressive on their own, neither alone is Iron Man.

Iron Man is the system.

The point of the analogy is not that Stark is augmented by a powerful tool. It is that Iron Man, the actual capability that walks out of the workshop and does things neither Stark nor Jarvis could do, exists only when the two operate as a single integrated unit. Stark's judgement, intuition, creativity and accountability are directing Jarvis's processing power, recall and execution capacity in real time, continuously, while the work is happening. Jarvis's contribution is not raw computation handed to Stark on request. It is contribution shaped by the goal of the moment, surfaced at the moment Stark needs it, in a form Stark can act on without breaking stride.

Watch the films closely and the interesting moments are not the ones where Stark issues a command and Jarvis carries it out. Those are the moments where the analogy collapses into Task Delegation. The interesting moments are the ones where Stark is mid-thought, Jarvis is surfacing the next consideration before Stark has finished articulating the current one and the resulting action is something Stark could not have arrived at without Jarvis and Jarvis could not have arrived at without Stark. Stark is not faster because of Jarvis. He is reaching better decisions, under pressure, against constraints he could not have held in mind alone.

That is the dynamic Paired Cognition is trying to produce. The fiction exaggerates the responsiveness, the seamlessness and the dramatic stakes, but the structural relationship between Stark and Jarvis is the part of the picture that survives transfer to a real workshop.

In professional contexts, the same dynamic produces strategy work, legal analysis, financial models and clinical reasoning that neither a senior practitioner nor an AI system could produce alone. The senior consultant becomes the Stark. The AI becomes the Jarvis. The partnership becomes capable of more than the sum of either part. And the gain shows up where it matters: in the quality of what is decided, not in how quickly the decision is reached.

The framing also clarifies what Paired Cognition is not. It is not Stark being replaced by Jarvis; that would be automation, and the resulting "Iron Man" would be a different and less interesting machine. It is not Jarvis acting without Stark's awareness; that would be Autonomous Agency, with all of its different governance demands. It is not Stark issuing orders, walking out of the workshop and reviewing reports; that would be Task Delegation, and the suit would never get built. The partnership is constant, the engagement is mutual and the resulting capability belongs to the system rather than to either component.

This is the image to hold throughout the rest of the book. When subsequent chapters refer to the human-AI partnership, the working unit or the joint cognitive system, they are pointing at the same thing. Iron Man is what happens when Stark and Jarvis are inseparable.

3.5 Directed execution is what produces qualitatively better outputs

There is a specific mechanism through which the shared mental model produces qualitatively different output, and it is worth naming now even though a later chapter develops it in depth. The mechanism is directed execution.

When a Paired Cognition partnership reaches the point of producing output, the AI is not working from a cold brief. It has participated in Exploration, Scoping and Planning. It understands why the work matters, who it serves, what constraints apply, what has already been considered and set aside and how this output fits into a larger picture. Execution becomes the natural continuation of a conversation that has been building context for hours or, sometimes, days.

This is what distinguishes directed execution from Task Delegation. The AI is not autonomous. It is also not narrowly instructed. It operates within a richly developed shared model, making local decisions that respect the broader frame because it has helped build the frame. The practitioner does not have to specify every detail, because the model itself constrains the space of acceptable outputs.

The contrast with Task Delegation is sharp. In Task Delegation, an AI receives a discrete instruction and executes it without surrounding context. The output is correct on its own terms but often misses the strategic frame. In directed execution, the same AI applied to the same nominal task produces work that fits the wider effort, because it has carried the context forward. A later chapter returns to this contrast with a side-by-side example that makes the gap visible at the level of the prose itself. For now, the structural point is enough.

This explains something organisations adopting AI find puzzling. The same tools, deployed against the same nominal tasks, produce wildly different results in different organisations. Although the capability of the model is similar in each case, the investment in shared context varies enormously. Organisations that build rich working models with their AI receive directed execution, and the output reflects it. Organisations that do not receive isolated task completion, no matter how capable the underlying model is. The difference compounds across every interaction in every workflow.

> *A wealth advisor preparing a recommendation for a long-standing client is not looking for a faster draft. She is working through a configuration of considerations her firm has not seen before: a recent inheritance, a family business in transition, a soft signal from the client that retirement is closer than the formal plan assumes.*
>
> *The AI in her partnership holds the firm's investment philosophy, the client's documented history, the regulatory posture across the relevant products and the soft notes from prior conversations. Rather than retrieving these as documents, it is reasoning with them as live considerations that shape the recommendation in real time. The recommendation that lands in the client's hands is one she could not have produced in the same time, and one a generic model could not have produced at all. The advisor decides. The partnership ensures she is deciding with everything in view.*

3.6 Paired Cognition fits where judgement is constant and context is dense

Paired Cognition is not the right model for every workflow. It is the right model for complex, contextual professional work in which judgement matters throughout and the cost of a generic output is high.

Strategy work, legal analysis, financial advisory, clinical reasoning, complex underwriting and senior product decisions all share this profile. They reward sustained collaboration because their value lies in nuance that no checklist can capture. The cost of getting them generically right but specifically wrong is significant, sometimes severe. The practitioner carries professional accountability that cannot be transferred to a machine. Paired Cognition fits these conditions because it strengthens the practitioner without displacing her.

Other workflows do not share this profile, and Paired Cognition would be wasted on them. A standard renewal letter does not require a shared mental model. A high-volume document classification task does not benefit from sustained dialogue. The other three operating models exist because there are large categories of legitimate AI work in which the partnership posture is the wrong posture.

This chapter's claim is narrower and stronger than a universal endorsement of partnership. It is that Paired Cognition exists, that it is the most under-recognised of the four operating models and that for a particular and economically significant class of work, it is the form of AI use that produces qualitatively better outcomes than anything else available. Where it fits, no amount of cleverer prompting inside a Task Delegation pattern will substitute for it. The partnership is the unit of work. The shared mental model is the mechanism. Directed execution is the form the output takes. The result is decision quality of a kind the dominant narrative around AI has, until now, been largely unable to describe.

The next chapter develops the other three operating models with the same care. Task Delegation, Autonomous Agency and Embedded Intelligence each carry their own logic, their own governance demands and their own ideal applications. None is a lesser cousin of partnership. Each is the right answer for some part of the work, and an organisation that learns to choose between them deliberately ends up far more capable than one that has only learned to partner, or only learned to delegate.

Iron Man is the system. The next chapter walks across the rest of the spectrum.

CHAPTER 4

Delegation, Agency and Embedding: When AI Works Without a Partner

> *Thick rules require experience and discretion to apply, and come with exceptions, examples, and provisos. Thin rules aspire to be self-sufficient and explicit and apply to all situations uniformly.*
>
> — Lorraine Daston

A great deal of valuable work does not need two minds in continuous dialogue. It needs execution. It needs supervision. It needs quiet automation behind the scenes. The previous chapter argued that Paired Cognition is the most under-recognised of the four operating models, and that argument stands. But treating Paired Cognition as the universal ideal would be a different kind of mistake, the mirror image of treating Task Delegation as the universal default.

Three further operating models cover the territory Paired Cognition does not. Each carries its own logic, its own governance demands and its own ideal applications. Each is the right answer for some part of an organisation's work. None is a lesser cousin of partnership.

The decisive question across all four models is the one introduced at the outset: where does decision-making responsibility sit, and at what stage of the work does it sit there? That single axis discriminates between the four. The answer changes everything else about how the work should be designed: who carries the cognitive load, where governance concentrates, what failure looks like and what good looks like.

4.1 Task Delegation is the model most organisations meet first

Task Delegation is the model most organisations meet on day one with AI, and the one many never move beyond. The pattern is simple. A human takes a body of work, decomposes it into discrete tasks and assigns those tasks to an AI for execution. The AI receives each task as a self-contained instruction. It returns a result. The human reviews, integrates and decides what comes next.

The defining feature of this model is that the AI does not share the broader context of the work. It is not building a mental model of the project alongside the human. It is performing isolated acts of execution against a brief.

> *A marketing manager preparing a product launch sends six prompts to an AI over the course of an afternoon. Three subject lines for the announcement email. A 200-word product description for the website. Five social posts in the brand's house tone. Each request stands alone.*
>
> *The AI does not know what the launch is for, what the brand is trying to achieve or how the pieces fit together. It does each task to a high standard and returns the result. The manager reads each output, accepts or revises it, and assembles the campaign in her head as she goes.*

Decision-making responsibility sits firmly with the human throughout. The human decides what to ask for, how to phrase the request, what to accept and how the output fits into the larger work. The AI is responsible only for execution within the scope of the task as specified. This division is not a limitation of the technology, but a deliberate property of the operating model.

Task Delegation works well when the work decomposes cleanly into well-understood units. A paralegal producing standard form letters from a matter file. An analyst asking for a regular expression that matches a known data pattern. A communications associate generating a clean first draft of a routine media statement. All of these benefit from fast, accurate execution without the overhead of a shared cognitive partnership. The work does not need a partner. It needs an executor.

4.2 Task Delegation carries a hidden cognitive cost

What Task Delegation gives away, almost invisibly, is the integration work. The AI never sees the surrounding shape of the project, so the human carries the entire job of holding the parts together in mind, deciding how each output fits and reconciling the small inconsistencies that accumulate when six discrete prompts are asked to behave as if they were one coherent piece of work. The cost is not in any single prompt. It is in the silent assembly that happens between prompts, inside the practitioner's head, and that no AI is helping with.

This shows up in two ways that are easy to miss. The first is the slow drift of tone and emphasis across a delegated body of work. The marketing manager's six outputs each meet the brief she gave, yet the launch as a whole reads as if it were written by someone whose attention kept moving, because in a sense it was. Reconciling those drifts at the end is its own task, often longer than the practitioner expected. The second is the loss of cognitive amplification on the harder parts of the work. Because the AI is never developing a model of the project, it is also never positioned to notice that two of the requests, taken together, point at a strategic question the practitioner has not yet asked. The opportunity to be challenged by the system is forfeited at the moment of decomposition.

The organisational consequence is that a practice which standardises on Task Delegation is, without quite intending to, paying the integration tax on every piece of work it ships, and forgoing the partnership benefit on the work where partnership would have changed the answer. The distinction between an AI working from a discrete brief and one working within a richly developed shared model is developed fully in a later chapter, including what that difference produces at the level of the output itself. The point worth carrying forward here is that Task Delegation is not free even when each individual delegation looks efficient.

4.3 Autonomous Agency moves the human into a supervisor's chair

Autonomous Agency shifts decision-making in a different direction. The human sets objectives, defines constraints and establishes guardrails, and the AI then works independently within that envelope, without each step being specified in advance. Within those bounds the AI determines its own path, makes intermediate decisions and reports back when the objective is met or when something falls outside the boundary.

The human role shifts from collaborator to supervisor, although that shift is anything but passive. The supervisory work is the act of designing the objectives and guardrails themselves, and that design becomes the primary governance act of the operating model.

> *A coding agent is assigned to fix a defined bug in a payments service. The on-call engineer describes the failing behaviour: duplicate ledger entries appearing under a specific retry condition. She points the agent at the relevant repository, the related test suite and the recent commit history, then steps away to attend to a separate incident.*
>
> *Over the next 40 minutes the agent reads the code paths involved, forms hypotheses about where the duplication originates, runs the existing tests to confirm the failure mode, writes a new test that isolates the bug, edits the retry logic and re-runs the suite until everything passes. It surfaces a single question along the way: a particular reconciliation behaviour could be tightened further, but only at the cost of a backwards-incompatible API change, which the agent flags rather than makes. The engineer returns to a green build, a written summary of the fix and one decision waiting for her judgement.*

Within the scope of that bug, the agent is making real decisions, not executing instructions. It chooses where to look. It chooses what hypothesis to test next. It chooses when its work is done. The engineer has not specified any of these steps in advance, and could not have, because she did not know which path through the code base would turn out to matter.

The pattern repeats across many domains.

> *An autonomous quality-control agent on a production line monitors sensor data continuously, flags anomalies that fall outside defined tolerances, halts the line when a flag exceeds a severity threshold and produces a structured report for the shift supervisor.*
>
> *A customer operations agent is authorised to resolve refund requests below a defined value when the request matches a known pattern, escalating any case that falls outside the pattern to a human team.*
>
> *A logistics agent reroutes deliveries around weather disruptions within strict bounds on cost increases and delivery windows, reporting its decisions for daily review.*

In each case, the AI holds decision-making responsibility within a bounded scope, while the human's job is to design the boundary. A poorly defined objective produces an agent that confidently solves the wrong problem. A poorly designed guardrail produces an agent that takes actions the organisation never intended to authorise.

4.4 Autonomy fits work that is bounded in consequence

Autonomous Agency suits work that is well-defined, repeatable and bounded in consequence. The clearer the objective, the more reliably an agent can pursue it. The more contained the blast radius of a wrong action, the more comfortable an organisation can be in granting the autonomy in the first place.

These two properties travel together in practice. Bug fixes inside a service with strong test coverage are bounded because the tests catch most of what could go wrong. Refund authorisations below a threshold are bounded because the dollar value caps the consequence of any single mistake. Reconciliation entries with hard limits on adjustment size are bounded because the limit constrains exposure even when the agent is wrong.

> *On the RE-Bench evaluation, the Stanford Institute for Human-Centered Artificial Intelligence (2025) reported that top AI systems outscored human experts roughly four to one when given a two-hour budget, but human experts outscored AI two to one once the budget extended to thirty-two hours: empirical evidence that autonomy degrades over longer horizons.*

Autonomy does not suit work where each instance is novel, where the cost of a wrong decision is severe or where the success criteria themselves are matters of judgement that shift as the work proceeds. A complex commercial negotiation is not a candidate for an autonomous agent. Neither is the framing of an organisation's response to a regulatory enquiry. Neither is a piece of strategic analysis where what counts as a good answer depends on an executive conversation that has not happened yet.

In those situations, the absence of a human partner during execution is a liability rather than a feature. One of the other operating models will fit better. Often it is Paired Cognition. Sometimes it is Task Delegation with sustained human oversight at every step. The judgement call is the design of the boundary itself.

4.5 Embedded Intelligence puts AI inside the plumbing

Embedded Intelligence is the operating model with the smallest human footprint at runtime. The AI is a component inside automated infrastructure. It is invoked by other software, processes inputs and produces outputs that flow on to the next stage of an automated pipeline. No human is in the loop while the work is happening. Gartner (2025) forecast that the share of enterprise applications shipping with task-specific AI components would rise from under 5 per cent that year to roughly 40 per cent in 2026. AI is increasingly inside the plumbing rather than at the surface where users meet it.

> *When a customer support ticket closes inside a software company's helpdesk system, an automated process sends the conversation transcript to an AI component that drafts a knowledge-base article, applies relevant tags and queues the draft for editorial review. The component runs hundreds of times a day without ever speaking to a person while it works.*
>
> *A human reviews the queue once or twice a week, accepting some drafts as written, editing others and discarding the rest. The component is doing what a deterministic function might once have done, with the flexibility only a language model brings to language work.*

The same pattern extends across many independent agents performing behind-the-scenes duties. Nightly classification of incoming bug reports into priority bands. Weekly summarisation of user feedback across support, sales and product channels. Generation of test data for staging environments. Categorisation of inbound documents into routing queues. None of these touch a human at runtime. They are part of the plumbing of the organisation, invisible until they fail.

Conceptually, Embedded Intelligence is the simplest of the four models. It is traditional software architecture in which AI components have replaced what would once have been hand-written logic, rules engines or brittle pattern-matching code. The human work happens at design time: choosing where to place these components, what they should do, how their outputs flow through the rest of the system and how the system should behave when a component returns something unexpected.

That last point is where Embedded Intelligence quietly differs from the deterministic code it replaces. A language model component will sometimes produce output that a hand-written function would not. The architecture has to absorb that variability gracefully. The design-time work is more demanding than it looks, even though the runtime work is invisible.

4.6 Most organisations will run all four models at once

The four operating models are not mutually exclusive, and most organisations of any size will use all of them. A single department often runs all four in parallel without the people involved noticing.

> *Inside the operations function of a mid-sized insurer, Embedded Intelligence runs in the document ingestion pipeline that extracts structured data from broker submissions overnight. Autonomous Agency handles the routine triage of low-value endorsement requests within strict bounds on policy change.*
>
> *Task Delegation produces the standard decline letters, renewal reminders and policy schedules that the team sends out by the hundreds each week. Paired Cognition is what the senior underwriter does when a complex industrial submission lands on her desk and she works through it with AI alongside her over the course of an afternoon. Each task finds the model that fits its shape.*

The fit is determined by two questions. Where does the cognitive load of the work belong: with the human, with the AI or shared between them? And where does decision-making responsibility need to sit, given the consequences of the decisions involved?

Workflows with high uncertainty, novel structure or stakes that resist specification belong with humans, often in partnership. Workflows with stable structure, clear specifications and bounded outcomes are candidates for delegation, agency or embedding, depending on how much autonomy is appropriate and whether a human interface is needed at all.

The choice is not about which model is best. It is about which model matches the shape of the work and the level of risk the organisation is prepared to absorb.

4.7 The real risk is Task Delegation becoming the unexamined default

Many organisations adopt Task Delegation first and stop there. The reasons are understandable. It is the easiest model to introduce. It fits cleanly into existing job descriptions. It requires the least change to how decisions are made. People can use AI as a faster execution engine without rethinking anything else about how they work.

The risk is not that Task Delegation is wrong. It is that Task Delegation becomes the unexamined default for every kind of AI work, including work whose shape calls for something else.

Work that would benefit from genuine cognitive partnership gets fragmented into discrete prompts, losing the continuity that makes Paired Cognition valuable. Work that could be safely automated through Embedded Intelligence stays manual, because no one notices that the human review step adds nothing meaningful. Work that could be handled by an autonomous agent stays in the inbox of a human who is no better placed to do it than the agent would be.

Each of the three models in this chapter solves a real problem. Although none of them looks like the partnership that the previous chapter celebrated, each is the right answer for some part of an organisation's work. Read as tools with their own characters rather than as compromised forms of partnership, they reveal the shape of the work they were chosen for.

An organisation that learns to recognise which model fits which workflow ends up with a far more capable operation than one that has only learned to delegate. That recognition is a discipline rather than an instinct. The next chapter develops it directly, in the form of six criteria for choosing a model, four diagnostic symptoms of mismatch and the sequence that puts model selection ahead of tool selection. The work of Part I closes with that selection question, because every later part of the book assumes the question has been answered.

Chapter 5

Choosing the Right Operating Model

I suppose it is tempting, if the only tool you have is a hammer, to treat everything as if it were a nail.

— Abraham Maslow

A senior partner at a mid-sized law firm spent six months convinced her practice had an AI problem. The firm had bought a sophisticated contract drafting platform. Associates dutifully fed it briefs. Outputs came back fluent and confidently wrong, and reviewers spent more time correcting drafts than they would have spent writing from scratch. Meanwhile, the same partner kept a personal AI session open on a second monitor and used it to think through every difficult matter that crossed her desk.

The technology, working exactly as designed, was not the source of the problem. The problem was that the firm had bought a Task Delegation tool and tried to do Paired Cognition work with it, while the partner, without anyone telling her to, had quietly assembled a Paired Cognition workflow on the side. The platform was solving the wrong problem in the right way. That is what a model mismatch looks like from inside an organisation.

This is the discipline at the centre of Part I. The four operating models are useful only if an organisation can choose between them deliberately, workflow by workflow, and resist the gravitational pull of whichever model its tooling, vendor or culture happens to prefer. Naming the four models is the easy part. Matching each one to the work that actually fits it is what separates organisations producing real outcomes from organisations producing expensive disappointment.

5.1 The wrong model imposes friction practitioners feel every day

A workflow that needs sustained partnership but has been forced into Task Delegation produces brittle outputs and demoralised reviewers. A workflow that needs background embedding but has been wrapped in a chat interface wastes attention on supervision that adds no value. The model and the work must match, or neither succeeds.

The previous chapters have built the vocabulary needed to make that match. Earlier chapters established that the four models differ along a single load-bearing axis: decision-making responsibility. They developed Paired Cognition as the model that produces quality gains rather than speed gains, and named it the most under-recognised of the four. They then worked through the other three, Task Delegation, Autonomous Agency and Embedded Intelligence, and warned against Task Delegation becoming the unexamined default.

What follows now is the decision discipline itself. How does an organisation, looking at a real workflow on a real Tuesday morning, decide which of the four models fits? Six criteria, taken together, do most of the work.

5.2 Six criteria, read as a conversation

The criteria function together, not in isolation. Read together, the answers cluster, and the cluster names the model.

Start with **complexity**. Work that involves many interacting variables, ambiguous goals or layered trade-offs tends to require sustained collaboration rather than discrete handoffs. Routine work with a small number of stable inputs and clear outputs is much easier to scope and delegate. Complexity is not the same as importance. A high-stakes workflow can still be simple, and a low-stakes workflow can still be complex. The relevant question is shape, not weight.

Once shape is on the table, the next question is where the shape comes from. That is **contextual dependency**. Much of what makes a workflow complex is not the work itself but the surrounding context it depends on: prior decisions, organisational history, the personalities of the people involved, the strategic posture of the firm. A document classifier inside an insurance pipeline does not need to know the politics of the underwriting committee. A board paper on market entry cannot proceed without that knowledge. Context is what makes complexity sticky.

Sticky context, in turn, is what creates a need for **sustained reasoning**. When the relevant material cannot be summarised into a brief, it has to be carried through the work itself, held in active memory, returned to as new threads emerge. This is the hallmark of cognitive partnership. Other problems break cleanly into independent steps that can be processed once and shipped. The shape of the reasoning indicates the shape of the appropriate operating model.

The mirror image of sustained reasoning is **specification clarity**. Some work can be described in advance with enough precision that an executor with no broader context can do it well: produce this document, summarise these inputs, classify these records. Other work resists such specification because the goal itself only becomes clear through engagement with the material. If the brief cannot be written until the work is half done, the workflow does not belong in Task Delegation. The criterion is not whether a brief is hard to write, but whether writing it is even possible before the work begins.

Specification leads naturally to a question about output. **Tolerance for AI variability** asks what happens when a generative system, run twice on identical inputs, produces meaningfully different results. While some workflows absorb that variability through downstream review or low-stakes use, others require deterministic behaviour, regulated outputs or auditability that cannot tolerate drift. Higher tolerance opens the door to more autonomous models. Lower tolerance pushes the work toward Paired Cognition or tightly bounded Embedded Intelligence.

The final criterion brings the human back into the frame. **The need for human judgement during execution** asks where, exactly, the judgement has to live. Some decisions must be made by a person at the moment they arise, not at design time. Others can be predetermined and encoded into rules or guardrails. Workflows that require live judgement during the work belong in models that keep a human in the loop. Workflows where judgement can be exercised once and then captured can move toward agency or embedding. Everything in the previous five criteria feeds into this last one, because together they decide whether the human's attention is needed in the room or only at the perimeter.

Read in series, the criteria do something a checklist cannot. They build a picture. Complex, context-heavy work that resists specification and demands live judgement clusters obviously toward Paired Cognition. Simple, well-scoped, variability-tolerant work with no live judgement requirement clusters toward Embedded Intelligence. The other two models occupy the middle, and the same chain of reasoning shows why: Task Delegation handles work that specifies cleanly but still needs human review of each output; Autonomous Agency handles work that runs within bounds but absorbs intermediate decisions the human does not need to make.

The six criteria are not a scoring rubric. They are the questions a team has to walk through in order, because the answer to each one shapes the next.

5.3 How does a mismatch announce itself?

A team that has chosen the wrong model rarely realises it directly. The diagnosis arrives indirectly, through symptoms.

The first symptom is **persistent friction**. Practitioners feel that the AI is in their way. Prompts go nowhere, outputs need to be redone, and ceremony accumulates around work that did not previously need any. The friction is not a sign that the practitioners are resistant to AI. It is a sign that the model is wrong for the work.

The second symptom is **quality shortfall.** Outputs are technically delivered but consistently fall short of what the work demands. Reviewers spend more time fixing AI outputs than the original task would have taken. This usually means complex, contextual work has been routed through Task Delegation when it actually needed Paired Cognition.

The third symptom is **over-automation of complex work**. An organisation builds an autonomous agent or an embedded pipeline for work that genuinely required human judgement at runtime. The system runs, produces plausible outputs, and erodes quality in ways that are hard to detect until something breaks publicly. Confidence in the output decouples from the actual reliability of the output. By the time the gap becomes visible, the workflow has been running on the wrong model for months.

The fourth symptom is the most diagnostic. **Practitioners route work around the system.** They use the official tool to satisfy a process requirement, then redo the substance somewhere else, often in a personal AI session that they actually trust. When skilled people quietly bypass the deployed solution, the deployed solution has been built on the wrong operating model. The MIT NANDA Initiative's *GenAI Divide* study (2025) found that more than 90 per cent of surveyed companies had employees regularly using personal AI tools for work, while only around 40 per cent had purchased a sanctioned enterprise LLM subscription: the empirical fingerprint of exactly this shadow workflow pattern.

That last pattern is worth pausing on, because it is the one most likely to be invisible to leadership. The official metrics still look healthy. The tool is being used, the process is being followed, the outputs are being produced. What is missing from the dashboard is the parallel workflow happening on a personal device, where the real thinking is done. The senior partner from the opening of this chapter was running exactly this pattern. So is at least one person in most organisations that have been deploying AI for more than a year.

Shadow workflows are the clearest signal a deployment can give, and they show up before the metrics do.

5.4 How four models look inside one organisation

Concrete examples make the framework usable. Consider a regional commercial bank with a credit team, an operations group, a treasury function and a core transaction platform. One organisation, four operating models, each chosen for the workflow rather than for corporate preference.

> *A senior credit officer is reviewing a borderline commercial lending proposal. The borrower has unusual cash flow characteristics, the security package is mixed, and the relationship history with the bank stretches back 15 years. The officer is not looking for a faster decision. She is looking for a partner who can hold the full credit policy, the bank's recent losses in this sector and the specific covenants that have worked on similar deals, all in active context as she thinks through the recommendation.*
>
> *This is Paired Cognition. The AI is in the room as the reasoning happens, surfacing precedents, testing assumptions and helping draft the credit memo as the shared mental model develops. The officer decides. The partnership ensures she is deciding with everything in view.*

The same bank's operations team faces a different shape of work. Each month, 20 branches produce variance commentaries on their performance against budget. The format is stable. The inputs are known. The financial controllers can specify exactly what is needed. This is Task Delegation. A clear brief, a generated draft per branch, a short review by each controller. The work is well-scoped and benefits from speed rather than partnership.

The treasury group runs intraday cash positioning across a dozen accounts. Within tightly defined parameters covering limits, counterparties and instrument types, an autonomous agent moves balances, executes routine transfers and rebalances exposures. A treasurer reviews the agent's actions at the end of each day and signs off the audit trail. This is Autonomous Agency. The AI decides within bounds the humans set in advance, and the humans supervise the pattern of decisions rather than each individual one.

Inside the core transaction platform, AI runs as Embedded Intelligence. Fraud signals on every payment. Identity matching during onboarding. Document understanding inside the loan processing pipeline. There is no chat interface, no prompt and no conversation. The intelligence is part of the plumbing, configured once and invoked at scale, with human review reserved for exceptions.

One bank, four models, each chosen because the work demanded it and not because the bank had a preferred way of doing things. The credit officer's deal could not be embedded; it requires live judgement and contextual reasoning that no design-time specification could capture. The fraud signals could not be routed through Paired Cognition; there are millions of payments and no time to think about each one. The variance commentaries could not be done by an autonomous agent; the controllers need to read the drafts, not delegate the responsibility for them. The treasury rebalancing could not be done as Paired Cognition; it would consume a treasurer's entire day to no purpose.

The same pattern appears in every sector. A professional services firm uses Paired Cognition for complex client matters, Task Delegation for routine document production, Autonomous Agency for portfolio monitoring across long-tail engagements and Embedded Intelligence for the matter management system that quietly classifies and routes incoming documents. A logistics business uses Paired Cognition for network design, Task Delegation for routine commercial correspondence, Autonomous Agency for dynamic rerouting within service-level constraints and Embedded Intelligence for the freight-matching engine. The shape repeats because the underlying logic is the same.

5.5 One model for everything: the most expensive failure

Organisations that adopt only one operating model pay a steep price. A firm that standardises on Paired Cognition for everything wastes expert attention on routine work that should have been delegated or embedded. A firm that standardises on Task Delegation never accesses the deeper value of partnership and treats AI as a slightly faster intern across every context. A firm that standardises on Autonomous Agency builds elaborate guardrails for work that did not need automation in the first place. A firm that standardises on Embedded Intelligence reduces AI to plumbing and never engages it as a thinking partner.

Each of these failures looks different from the inside. They are the same failure.

Monoculture usually emerges from one of two sources. The first is procurement. The organisation has bought a tool that supports one model, and every workflow gets reshaped to fit that tool. The second is culture. A leadership preference for automation, or a practitioner preference for partnership, becomes the default lens through which every adoption decision is filtered. Both patterns produce the same outcome: a portfolio of workflows with the wrong operating model on most of them.

The defence against monoculture is the framework itself. When the operating model is named explicitly for each workflow, and the selection is justified against the six criteria, it becomes much harder for a single preference to colonise the whole organisation. The discipline of naming forces the conversation that procurement and culture would otherwise short-circuit.

5.6 Model selection has to come before tool selection

A practical implication follows directly. The operating model decision must come before the tooling decision, because tools are designed around operating models even when their vendors do not say so. A conversational assistant assumes Paired Cognition. A workflow automation platform assumes Embedded Intelligence. An agentic framework assumes Autonomous Agency. A prompt-based productivity tool assumes Task Delegation. Choosing the tool first quietly chooses the model, often without anyone noticing.

The healthier sequence is the reverse. Identify the workflow. Read it against the six criteria. Name the operating model. Then ask which tooling fits. This sequence prevents the most expensive adoption mistake an organisation can make, which is buying a platform and discovering 18 months later that the workflow it serves best is not the workflow that mattered most.

Model first. Tool second. Every other order leaks value.

5.7 The selection mistakes that show up most often

Several patterns repeat across organisations. The first is treating Paired Cognition as a luxury rather than as the right model for genuinely complex work. Organisations describe partnership as expensive or slow, then watch their delegated outputs underperform on exactly the work that needed it. The cost they avoided was lower than the cost they incurred by avoiding it.

The second is treating Embedded Intelligence as an aspiration for every workflow, on the assumption that anything not yet automated is simply behind schedule. Some workflows should never be embedded, because their value depends on human judgement at runtime. To embed those workflows, far from being progress, destroys the thing that made them valuable in the first place.

The third mistake is conflating Task Delegation with Autonomous Agency. They differ on a critical axis: who decides what to do next. Where delegation has the human decide and the AI execute, agency lets the AI decide within bounds the humans have set. The governance, monitoring and risk profile differ accordingly. Treating them as interchangeable produces either over-supervised agents or under-supervised delegations, and both fail in expensive ways.

The fourth mistake is changing the model implicitly when the work changes. A workflow begins as Paired Cognition during design, becomes Task Delegation during routine use and quietly becomes Autonomous Agency once the team trusts the system. Each of those transitions can be legitimate, provided someone has actually decided to make it. The danger is the drift that no one names: an organisation ends up with autonomous systems doing work that no human ever signed off on, because no one was asked to.

The common thread across all four is the same. In every case, the model is being treated as a property of something other than the workflow: the tool that has been bought, or the team's preferred way of operating. Every time the framing shifts away from the work itself, the wrong model gets chosen.

5.8 What Part I has built, and what Part II opens

Part I has done what it set out to do. It has named the four operating models, distinguished them along the axis of decision-making responsibility, developed Paired Cognition as the most under-recognised of the four and offered the discipline by which an organisation chooses between them workflow by workflow. The reader who has reached this page has the vocabulary and the selection method that the rest of the book assumes.

Part II opens the engine of Paired Cognition. If the model has been named correctly for a workflow, the next questions are operational. How do a human and an AI actually maintain a shared mental model across a piece of work? What is the cognitive architecture beneath the partnership? Where does organisational memory live, and how does it accumulate? What conversation types are available, and what work phases does Paired Cognition pass through? What does it mean to direct execution rather than delegate it?

These questions only become tractable once the model has been chosen well. An organisation that has not yet learned to select between the four models cannot operate Paired Cognition meaningfully, because every Paired Cognition workflow will be in competition with the gravitational pull of the wrong tool, the wrong team configuration or the wrong default assumption about who is responsible for which decisions. Part II is the reward for getting the selection right.

Selection alone is necessary but not sufficient. An organisation with 20 workflows and four operating models faces a further problem: how to compose those models so work flows cleanly across them, governance remains coherent and the boundaries between models stay explicit. A document might begin inside an Embedded Intelligence pipeline, surface into a Task Delegation queue and end as input to a Paired Cognition strategy session. The handoffs matter as much as the individual choices. That composition problem, how work flows across models, is treated as a discipline in its own right later in the book.

For now, the foundation is the per-workflow decision. Get that right, repeatedly, across the work that actually matters, and everything that follows has somewhere coherent to land. Part I has built the four-model framework and the discipline that selects between the models; Part II opens the most demanding of the four and walks through the architecture that makes it work.

Part II

Inside Paired Cognition

The cognitive architecture of partnership: externalised thinking, the three-node model, the knowledge base, conversation types, the seven workflow phases, and the discipline of directed execution.

Chapter 6

Externalised Cognition: Thinking Out Loud With AI

How can I tell what I think till I see what I say?

— E. M. Forster

A senior tax partner can see the shape of a complex restructure within 10 minutes of opening the file. Producing the written advice for the client takes the rest of the week. The expertise is fast while the expression is slow, and most knowledge work runs on this gap, with most of its inefficiency hiding inside it.

Paired Cognition rests on a single mechanism that closes that gap. A practitioner thinks out loud. An AI listens, structures what it has heard and returns something coherent. That structured output becomes the substrate for the next round of thinking. The loop, repeated and refined, is the operational engine of the partnership model. Without it, Paired Cognition is a concept. With it, the concept becomes a working practice.

Part II is about the inside of that practice. This chapter takes the loop apart. It explains why expressing unstructured thought is cognitively natural, why modern AI is unusually well suited to receiving such input and why the value lies not in the act of speaking but in the complete cycle the speaking initiates.

6.1 Why does professional life fight the way the mind actually works?

In *The Principles of Psychology*, published in 1890, William James described consciousness not as a sequence of discrete ideas but as a continuous stream that flows, branches and doubles back on itself. In his account, thought arrives in fragments, associations, half-formed comparisons and sudden reversals, never in tidy sentences. The mind moves faster than language, and faster still than the formal structures that language is later forced to wear.

Professional environments demand the opposite. A board paper requires a clear thesis. A regulatory submission requires a defensible structure. An architectural decision record requires precision and brevity. The work of converting raw cognition into a finished artefact is substantial, and for many practitioners it consumes more time than the underlying thinking it represents.

The expert often knows the answer long before they can write it down.

This is the gap where most knowledge work loses its leverage. A credit officer can identify the structural weakness in a borrower's proposal in the time it takes to scan the term sheet. The written submission takes days. A campaign strategist can feel the shape of a positioning argument almost immediately. Documenting the reasoning for a client deck takes far longer. The gap between knowing and articulating is not a failure of the practitioner. It is the standing tax that traditional knowledge work pays on every output it produces.

Closing that gap is what Externalised Cognition is for.

6.2 Externalisation suspends the editor and lets thinking run at its natural pace

Externalisation is the act of expressing raw cognitive content through speech or text without pre-structuring or filtering. The practitioner permits themselves to think aloud rather than to compose. Headings, transitions and tidy ordering are deferred. What comes out first is not the artefact, but the thinking that an artefact would normally have to be carved from.

For experienced professionals, this can feel uncomfortable at the start. Most have spent careers learning to present polished output and to suppress whatever did not survive the polishing. Speaking before structuring violates a deep professional instinct. The friction of pre-structuring is precisely what externalisation removes. By suspending the editor, the practitioner gains access to the full bandwidth of their thinking, including the lateral associations and tacit knowledge that rarely survive formal drafting.

Speech-to-text is the practical facilitator that makes this routine rather than effortful. A practitioner speaks naturally, often while walking, driving or pacing in an office, and the spoken content is captured as text. The keyboard, with its enforced linearity and its slower clock speed, falls away. The pace of speech, which is closer to the pace of thought, becomes the pace of capture.

A 15-minute monologue can produce 2,000 words of raw cognitive content that would otherwise have taken hours to type, if it could have been typed at all. A controlled Stanford HCI study by Ruan and colleagues (2017) measured speech recognition at roughly three times the speed of a state-of-the-art smartphone keyboard for English, and 2.8 times for Mandarin, with lower error rates than typing: the empirical anchor for what the keyboard's enforced linearity costs the speaker who already knows what they think.

> *An advertising strategist preparing a positioning recommendation for a challenger brand spends 25 minutes walking around her office speaking through the brief. She moves between the category dynamics, the brand's actual strengths versus the strengths it claims, the cultural moment the campaign would land in, the three positioning territories she had ruled out and why and the one she keeps coming back to.*
>
> *None of this is in order. Some of it contradicts other parts of it. She does not pause to fix any of it. The transcript will be 3,600 words long. It is the first time she has captured the full shape of her thinking in a single pass.*

6.3 A transcript is not yet useful

Externalisation alone produces a record. A record, however valuable, is not yet an artefact the practitioner can work with. The transcript contains repetition, false starts, half-finished sentences and ideas raised in one order that belong in another. Read in isolation, stream of consciousness output is often less coherent than the thinking behind it.

This is where the second half of the loop matters.

The AI ingests the externalised content and structures it. It identifies the underlying argument, groups related ideas, removes redundancy and produces an organised draft: a memo, a brief, a decision log, a positioning rationale. The structured output is not the end of the process. It feeds back into the shared context between practitioner and AI and becomes the basis for the next round of thinking.

The term Externalised Cognition refers to this complete loop, not to externalisation alone. Externalisation is the speech act. Externalised Cognition is the cycle the speech act sets in motion: human externalises, AI structures, human reviews and reacts, AI restructures. Because the value lives in the cycle rather than in either half, each iteration tightens the artefact and clarifies the thinking behind it. The output is co-produced. The cognitive load is genuinely shared.

The distinction matters because it explains why earlier attempts to harness AI for professional drafting fell short. Submitting a polished prompt and receiving a polished response is a useful transaction, yet it leaves the practitioner's thinking unchanged. The loop, by contrast, becomes part of the thinking itself, so the artefact is shaped during cognition, not after.

6.4 Modern AI is unusually suited to the specific work of structuring raw thought

Receiving stream of consciousness input and producing structured output is, on the face of it, a difficult task. It requires holding a large and disorganised body of material in working memory, identifying its underlying logic and rearranging it without distorting the speaker's intent. Humans can do this for short passages, but it is laborious. Most professionals delegate it reluctantly even to trusted colleagues, and only on work where the stakes justify the imposition.

Modern AI is unusually effective at exactly this operation.

It can absorb a long transcript in a single pass, identify recurring themes, reconcile inconsistencies and produce a structured rendering that preserves the speaker's voice and reasoning. It does not tire across the third or fourth iteration. It does not impose its own narrative preferences on the material the way a human ghostwriter sometimes will. It treats the practitioner's wording as the source of truth and arranges around it rather than over it.

Far from being a marginal convenience, this capability is the specific affordance that makes Paired Cognition practical. Without an entity capable of receiving raw cognitive output and returning structured content reliably and quickly, the partnership model collapses back into the older pattern of the expert producing finished work alone, with all the drafting tax that pattern carries. The loop closes only because AI can hold up its end of it.

A second affordance matters less obviously. The AI does not require the practitioner to know in advance what they are going to say. It accepts the thinking as it actually is, incomplete and recursive and occasionally self-contradicting, and renders it usefully on the first pass. The practitioner is freed from the discipline of knowing their argument before they have made it.

This is what makes the loop generative rather than merely efficient. A practitioner who has to know their argument in advance can only ever produce arguments they already had. A practitioner who can externalise and read the structured rendering back can discover, in the structuring, what they actually thought. Some of the strongest output produced through Externalised Cognition is content the practitioner could not have written from a blank page because they had not yet realised they held the view in question.

6.5 The same loop, across very different desks

The mechanism looks much the same whether the practitioner is preparing a credit submission, a software architecture review or a recommendation to a client board. What varies is the shape of the artefact at the other end and the conventions it has to honour.

> *A software architect at a payments company is weighing whether to split a monolithic settlement service into two domain-aligned services or to introduce a queue-based seam inside the existing service. He spends half an hour speaking through the decision: the latency budget, the failure modes the team has actually seen in production, the operational cost of running another service, the migration risk against the current release pressure, the position the platform team has taken on similar splits.*
>
> *The AI returns an architectural decision record with the decision, the alternatives considered, the consequences and the constraints. The architect reads it, realises he has not given enough weight to one of the failure modes, and speaks again. The next version reflects the new emphasis. Two more passes and the record is ready for review.*

> *A communications director responsible for a regulator-facing announcement at a listed company sits with the draft data, the legal team's redlines, the CEO's instinct that the framing is too defensive and her own sense of where the analyst community will press hardest. She speaks for 20 minutes, moving across all of it.*
>
> *The AI produces a structured brief: the core message, the supporting points, the questions she expects to receive, the lines she would refuse to give. She rejects the framing in the first version, says why, and externalises again. The brief*

converges over an afternoon. The thinking it captures would normally have lived in her head until the morning of the announcement.

A management consultant preparing recommendations for a private-equity-backed manufacturer's board does not begin with a slide outline. She spends 30 minutes speaking through the situation: the operating model the sponsor inherited, the three strategic options the team has worked through, the risks attached to each, her own intuitions about which path is defensible at the next valuation event.

The AI returns a structured argument with options, trade-offs and a recommended path. She refines by speaking again, this time reacting to what she now sees on the page in front of her.

The common pattern is visible across all of these contexts. The expert thinks. The thinking is captured. The capture is structured. The structure is reviewed. The review produces more thinking. At no point does the expert sit and write a document in the traditional sense. The document emerges from the loop.

What also varies, usefully, is what the practitioner notices when they read the structured output back. The architect realised mid-loop that he had been underweighting a known failure mode. The communications director rejected the framing only after seeing it laid out. The consultant clarified her recommendation by reacting to the version the AI had produced from her own words. In each case the structured output served less as a draft than as a mirror. The thinking became visible to the thinker, and the thinker thought again.

This is the part of the loop that practitioners new to Paired Cognition tend to underestimate. The instinct is to treat the AI's first structured output as the work, to be edited or signed off. The practice is to treat it as input: a high-fidelity rendering of one's own thinking that exists primarily to be reacted to. The third or fourth iteration is usually where the artefact becomes worth keeping. The first is rarely the destination.

6.6 What a single turn of the loop sounds like

Described abstractly, the loop can sound tidier than it is in practice. The following extract is an illustrative example, stylised for compactness rather than transcribed from any one session, of what a single turn of externalisation and restructuring sounds like when a practitioner is working at pace. The reader will recognise the texture even where the specific words are not theirs.

> Practitioner, speaking: *"Right, so the question is whether we go ahead with the regional rollout in October or hold it to February. October gets us in front of the buying season, but the operations team is already stretched on the integration, and Mei flagged last week that the support headcount is short for the volumes we'd see. Counter to that, every month we delay we lose ground to the incumbent, and the board paper in November will read very differently if we have early traction to point to. February is safer, but safer for whom. The CEO has already told the chair we'd be live this year. Park that. The real question is whether the operations risk is the kind that breaks the rollout or the kind that bruises it."*
>
> AI, returning a first restructuring: *"The decision sits between two timing options, October and February, with three considerations weighing against each other. In favour of October: market positioning ahead of the buying season, board narrative in November, and a prior commitment from the CEO to the chair. Against October: stretched operations capacity, a flagged shortfall in support headcount and the risk of a degraded launch experience. The framing question you ended on, whether the operations risk would break the rollout or bruise it, looks like the load-bearing one and is not yet answered."*
>
> Practitioner, pushing back: *"That's mostly right, but you've ranked the CEO's commitment too high. It is not a reason, it is a constraint I'd rather argue with than honour. Drop it down the list. And the bruise-versus-break question is the only one that matters. Take the support headcount issue and treat it as a probability, not a fact. What would have to be true for it to break the rollout, and what would have to be true for it to merely bruise it. Restructure around that."*

What this short extract shows is the asymmetry the chapter has been describing. The practitioner's first pass is not an argument, it is a set of considerations in a half-ordered tumble. The AI's first structured rendering is recognisably the practitioner's thinking, returned in a form she can react to. The reaction is where the work moves forward, because she can now see what she had been weighing, disagree with the weighting and redirect the next iteration to the question she actually wants pressure on. None of the three turns is the finished artefact. The finished artefact is several turns away, and each turn sharpens the underlying thought.

6.7 The cognitive division of labour matches the actual strengths of each party

The case for Externalised Cognition becomes clearer when set against the alternative. A partnership model that required experts to produce structured prompts, to format their thinking carefully before submitting it to the AI, would simply replace one drafting burden with another. The expert would spend their time preparing input rather than preparing output. The gain would be modest, and most practitioners would, after a few attempts, return to writing the thing themselves.

The shift to externalisation removes that burden entirely.

The expert is permitted to think the way they naturally think, in fragments and associations and half-finished arguments, and the AI assumes responsibility for converting that material into something usable. The division of labour is matched to the actual strengths of each party. Humans bring judgement, lateral thinking, domain expertise and the capacity to recognise what matters. AI brings linguistic capability, structural fluency and tireless reorganisation.

This division is what makes the model practical at the scale of an organisation. An institution that adopts Externalised Cognition as a working practice does not need to retrain its experts to be better writers. It needs to give them the means to externalise and the workflows that close the loop. The expertise that was previously trapped behind the bottleneck of formal drafting becomes accessible, repeatable and shareable across the firm.

A cultural implication accompanies the practical one. Once externalisation becomes routine, the markers of seniority shift. The practitioner who can produce a tidy document under deadline is no longer doing rare work. The practitioner who can think clearly out loud, who has something genuinely worth structuring when the AI receives it, is. The bottleneck moves from the keyboard back to the mind.

6.8 The mechanism is simple, and that is precisely why it scales

Externalised Cognition is not a novel theory of intelligence. It is not a new tool category. It is a working practice built on a particular kind of conversation between a person who knows something and a system that can shape what they say into something the rest of the organisation can use.

The mechanism is simple. The implications are not.

It is the foundation on which the rest of Part II is built. The next chapter introduces the second cognitive structure that Paired Cognition relies on, the three-node model in which human, AI and a knowledge base form a distributed cognitive system, and shows how the loop described here extends from a private practice into a team and an organisation. The chapter after that develops the third node, the knowledge base, in detail. Together they describe the architecture inside which thinking out loud becomes more than a personal technique and starts to compound.

CHAPTER 7

Distributed Cognition: The Three-Node Model of Human-AI Teams

The power of the unaided mind is highly overrated. Without external aids, memory, thought, and reasoning are all constrained.

— Donald Norman

In 1995, the cognitive anthropologist Edwin Hutchins published a study of navigation aboard a US Navy ship coming into San Diego harbour. The finding ran against intuition then and still does. No single sailor on the bridge knew how to navigate the vessel. The capability lived in the system as a whole, distributed across charts, instruments, communication protocols and a handful of crew members coordinating in real time. The pelorus operator called bearings to the bearing taker, who relayed them to the plotter, who fixed the ship's position on the chart. Each held a fragment. None held the whole. The ship found its way because cognition was shared between minds and artefacts.

Hutchins called this distributed cognition. The framing has since become foundational in the study of expert work, and it overturns a quiet assumption most organisations still operate on: that thinking is something which happens behind a single forehead, and that the artefacts surrounding the thinker are merely props.

They are not props. They are participants.

7.1 Expertise is a configured thing, not a property of one mind

Take the assumption seriously and an everyday observation begins to look strange. The senior litigator does not carry every precedent in working memory. They rely on a research librarian, a precedent bank, a junior associate's memo and a structured matter file. The clinician does not hold every protocol in their head.

They reach for guidelines, decision aids and the muscle memory of a team that has worked the same wards for years. The investment analyst does not remember every covenant. They keep a model, a deal book and a desk full of Bloomberg windows.

Each of these professionals is genuinely expert. The expertise is also configured. Remove the configuration and the expertise still exists, but its expression slows dramatically. Take the precedent bank away from the litigator and ask them to draft from memory, and the work is recognisably worse. Not because they have become less capable overnight, but because half their thinking apparatus has been packed into a box.

Configuration of this kind is the structure of professional work, not a limitation imposed on it.

Paired Cognition takes Hutchins seriously and pushes the implication one step further. If cognition is already distributed, then the relevant question is not whether to distribute it. The relevant question is how to design the distribution deliberately, so that the right capability lives in the right place, and the seams between human, machine and memory hold under load. The three-node model is the answer this book proposes.

7.2 Three nodes that each contribute something the others cannot

The first node is the human. The human contributes domain expertise built over years of practice, lateral thinking and the kind of reasoning that draws on lived experience. They bring judgement: the capacity to weigh competing considerations against an intuition for what matters in this particular case. They notice what is unusual. They sense when something is wrong before they can articulate why. They carry the relationships, the accountability and the professional standing that the work ultimately rests on.

The second node is the AI. It contributes a different and complementary capability set: fluent language, broad cross-domain knowledge and the ability to hold large amounts of context coherently within a single working session. It can structure unstructured input, follow chains of reasoning across many steps and produce artefacts at speed. It does not tire, does not lose patience and does not skip the boring parts.

The third node is the knowledge base. This is the canonical, retrievable memory of the organisation. It holds prior decisions, internal conventions, domain knowledge, regulatory interpretations and institutional context: the things a competent senior practitioner knows that a competent stranger does not.

The knowledge base is not a passive document store. It is structured for cognitive access during work, written in a form an AI can retrieve and reason over in the flow of a conversation rather than as a separate research task that must be initiated and then waited on. The next chapter develops this third node in detail. For the architecture, what matters is that it is present at all.

These three nodes are not interchangeable. Reasoning happens between the human and the AI, while the knowledge base holds rather than thinks, and Paired Cognition is the disciplined coordination of all three during active work.

7.3 The cognitive function and the sources it draws upon

A distinction is useful here, because it is the one most often blurred when people first encounter the model.

The cognitive function is the active thinking. It is the work of reasoning happening in real time, provided jointly by the human and the AI together. They are the two parties producing thought during the session, exchanging language, refining ideas, considering alternatives, converging on outputs. When a partner and an AI work through the structure of a deal aloud, the cognitive function is what is happening between them.

The sources are different. The sources are what the cognitive function reaches into as it does its work. The human's mind is one source: tacit and experiential knowledge absorbed across a career, much of it not consciously available until a relevant cue surfaces it. The knowledge base is the other: the explicit and codified knowledge the organisation has accumulated collectively. Both are queried during the session and both shape the trajectory of the work, although neither is doing the active reasoning.

Cognitive function is what is done. Sources are what is drawn upon.

The line matters because it clarifies what the knowledge base is for. It is not a replacement for the human's mind, and it is not a search engine bolted on the side of the conversation. It is one of two reservoirs the cognitive function draws upon. When a partner is drafting advice, the knowledge base is consulted in the same breath as the partner's own recollection, inside the act of thinking rather than as a separate research errand.

The line also clarifies what the AI is and is not: one half of the cognitive function, alongside the human. Although it has access to the knowledge base, the AI is not itself the knowledge base, which remains a structured artefact rather than a participant. Confusing these roles is the most common architectural error organisations make in their first year of Paired Cognition work. They treat the AI as either a tool to query or an oracle to consult, when in practice it functions as a thinking partner with shared access to the organisation's memory.

7.4 How does it look on a live matter?

> *A mid-sized commercial law firm is advising on a cross-border acquisition. A senior partner leads the matter. In a traditional setup, she relies on her own memory of similar deals, a junior associate to research precedents and a shared drive of past memos that no one has time to open. The work is high quality. The bottleneck is her calendar.*
>
> *In a Paired Cognition setup, the same partner externalises her thinking aloud: the buyer's strategic position, the unusual structure of the earn-out, the regulatory wrinkle in the target's home jurisdiction, the bits of the target's disclosure that do not feel right. The AI structures what she is saying, drafts the framework of the advice and proposes the lines of analysis the matter requires.*
>
> *As the conversation unfolds, the AI draws on the firm's knowledge base. The base holds the firm's house position on warranty caps, its standard escrow structures and the partner's own prior reasoning on a 2023 deal with a similar earn-out shape. The clause she would otherwise have to remember is surfaced into the conversation, with its original context attached. She decides whether it applies. The cognitive function, partner plus AI, moves faster and reasons more broadly because both sources are open at once.*

The partner is no longer the sole repository of the firm's accumulated judgement. Her own intuition still drives the work. It is supplemented in real time by the firm's collective intelligence, made retrievable through the knowledge base and brought into the cognitive function by the AI. The partner's reflective scrutiny of what the AI surfaces is itself the active part of her cognition, not a formality.

Chi and colleagues (1994) showed in their *Cognitive Science* study that students prompted to self-explain learned with substantially greater understanding than those who simply re-read the same material; self-explanation, they argued, is constructive, integrative and error-correcting in a way passive review is not. The partner who interrogates an AI-surfaced clause, asking why it applies and where it might not, is doing the same constructive work on a live matter.

The same architecture appears outside law. A wealth advisory firm encodes its investment philosophy, client segmentation rules and prior portfolio decisions into a knowledge base. An advisor working with AI produces client-specific recommendations grounded in the firm's house view rather than in the advisor's personal preferences. A clinical service builds a knowledge base of treatment protocols, departmental conventions and prior case formulations, and the duty consultant brings this canonical memory into the room during a complex admission. The shape of the architecture is the same. The contents of the third node differ.

It is worth pausing on what is genuinely new in this picture and what is not. Firms have always kept precedent banks. Hospitals have always kept guidelines. Investment houses have always kept a house view. The novelty is not the existence of organisational memory. The novelty is that the memory is now retrievable inside the cognitive function, by the AI half of it, in the moment the work is being done. The friction that used to separate stored knowledge from active reasoning has collapsed. What used to be a research errand is now a property of the conversation.

That collapse is where the gain comes from, and it is also where the design challenge sits. A knowledge base that the AI cannot retrieve from is not part of the architecture. It is still a repository, sitting outside the cognitive function as it has always done. The third node only becomes a node when it is reachable in the flow of the work.

7.5 Only one of the three nodes compounds, and it changes the others

The three nodes do not improve at the same rate.

The human node improves slowly, through experience, training and deliberate practice over years. The AI node improves through model upgrades and refined ways of working: gains that are real but largely outside the organisation's control. The knowledge base, by contrast, can compound. Every working session adds to it. Every decision recorded becomes a precedent. Every refinement in the firm's position becomes a clarification in the canonical memory.

As the knowledge base grows, the dependency profile of the organisation shifts. Decisions begin to draw on accumulated organisational intelligence rather than the recollection of whichever individual happens to be in the room. The quality of reasoning rises because the inputs the cognitive function can reach are richer, more consistent and less subject to the gaps in any single person's memory.

This shift changes what it means to be a new team member. In a traditional model, a new associate or a new analyst spends their first two or three years primarily acquiring contextual knowledge. They are not yet useful for creative or strategic work, because they lack the context to know what is normal and what is anomalous. They contribute by absorbing rather than by producing.

In a Paired Cognition setup, the knowledge base carries much of the contextual load. The new joiner can contribute creativity, fresh perspective and lateral thinking from early in their tenure, because the canonical memory is available to them through the AI in the same way it is available to a partner. The trajectory of the early career changes. Less of it is spent learning the firm's history. More of it is spent applying the joiner's own thinking to live work, with the firm's history loaded into the conversation alongside them.

7.6 A firm protected, a senior released

Continuity is the first benefit, and the easiest to underestimate.

In a traditional firm, a senior partner who moves to a competitor takes a substantial portion of the firm's institutional knowledge with them. Years of nuanced positioning leave with the person. In a firm that has invested in its knowledge base, the canonical memory persists. The departing partner's contribution was real, and the loss is real, but the knowledge base is no longer the firm's only copy. The remaining team can keep reasoning at the level the firm has set, because the source material the cognitive function draws upon has not walked out of the building.

This is not about replacing people. It is about removing the fragility that comes from concentrating organisational memory inside a single mind. The organisation becomes less brittle. The individual becomes less burdened.

The human consequences are easy to miss but significant. Senior practitioners are often on call because they are the only point of access to certain knowledge. A consultant takes calls on weekends because nobody else remembers a particular case. A partner reads emails on holiday because a junior cannot answer a client question without her. A clinical lead is paged on leave because a registrar needs a steer that only she can give. The knowledge base, properly built, reduces this dependence.

Less dependence supports better work. It also supports better careers, because experts are no longer trapped by their own irreplaceability. They can take on new work, mentor more deliberately or take a genuine break, knowing that the system around them carries enough of the context to function in their absence. The architecture pays out in both organisational resilience and individual freedom.

There is a quieter benefit too, which leaders sometimes notice before they can articulate it. When the cognitive function has reliable access to a well-built knowledge base, senior practitioners spend less of their day repeating themselves. The same explanation given for the third time to the third junior in the third week becomes something the AI can offer, drawing on the partner's own previous explanations recorded in the knowledge base. The senior's attention is freed for work that genuinely needs it. The junior's question is answered well, and at the time it is asked. Both parties are better served than under a model in which all knowledge transfer routes through the senior in person.

7.7 What this looks like in practice today

It is worth giving the architecture a touch of physical reality, even at the cost of dating the description. In most working deployments today, the knowledge base lives as a collection of structured notes, often plain markdown or a similar lightweight text format, organised so that each entry is short enough to retrieve as a fragment rather than as a full document. The tool families in use range from note-taking applications and team wikis through to dedicated knowledge-base products and vector-store-backed retrieval layers wired into the AI through whatever connector pattern the model supports at the time. Some teams keep the base in a code repository alongside the work it supports. Others use a hosted system with built-in retrieval. The specifics will date faster than this book will, and the named categories above will not be the right categories within a few years. The principle that survives the churn is the three-node distribution itself: a human, an AI and an externalised, retrievable memory, configured so that all three are reachable inside the cognitive function rather than around it.

7.8 The three-node model gives augmented intelligence a concrete shape

Paired Cognition is not an isolated proposal. It sits within a broader conversation about human-AI teaming that has been led, among others, by Microsoft Research and the Stanford Institute for Human-Centered Artificial Intelligence. Both research communities have argued, from different angles, that the productive frontier of AI lies in collaboration rather than substitution: in augmentation rather than replacement.

The augmented intelligence framing positions AI as amplifying human cognition rather than displacing it. The human remains the source of judgement, accountability and creative direction, while the AI extends that human's reach, speed and consistency. The three-node model gives this framing a concrete architecture. It names the knowledge base as the third element that turns episodic collaboration into durable organisational capability.

Without the knowledge base, every working session begins from scratch. The cognitive function does its work, the work concludes, and most of what was learned dissipates into a document on someone's drive. With the knowledge base, the work compounds. Each session adds to the source the next session will draw upon.

Seen from this angle, the move to Paired Cognition is less a technology adoption than an act of organisational design. The technology has matured to a point where the design becomes possible. The remaining work is largely human and structural: deciding how the three nodes connect, how the knowledge base is built and maintained and how teams adapt their habits to operate inside the new geometry.

The third node is where most organisations will discover whether they have built an architecture or a slide. Knowledge bases are easy to describe and hard to build well. The next chapter takes them up directly, and shows why a knowledge base is a different artefact from the document repositories most organisations already have.

Chapter 8

Building Organisational Memory: The Knowledge Base

The city does not tell its past, but contains it like the lines of a hand, written in the corners of the streets, the gratings of the windows, the banisters of the steps.

— Italo Calvino

Most organisations remember badly. They remember in inboxes and in the heads of senior people, in slide decks last opened in 2022, in folders named after projects no-one can recall the purpose of. They remember by accident more often than by design. And then, with depressing regularity, they forget: quietly, expensively and usually at the worst possible moment.

This is not a problem AI solves on its own. A capable language model brings extraordinary general knowledge to a conversation, but it does not arrive knowing why a particular team chose one approach over another in 2023, or what the regulator said in a meeting last spring or which clauses the firm has decided it will not negotiate. That knowledge belongs to the organisation. If it is not held somewhere both partners in the work can reach, it is not really held at all.

The knowledge base is the answer. It is the third node in the distributed cognition model introduced in the previous chapter, and it is the single most underrated investment a knowledge-intensive organisation can make in the era of capable AI. This chapter argues that case in detail.

8.1 Without a third node, the partnership stays thin

The previous chapter set out the three-node model: a human partner, an AI partner and a knowledge base. It also drew a distinction worth restating before going further. The cognitive function is the active thinking that human and AI perform together in real time. The human brain and the knowledge base are the sources from which that thinking draws. One is the work; the others are what the work is made of.

Without the knowledge base, every conversation starts from a standing start. The McKinsey Global Institute's 2012 *Social Economy* report (McKinsey Global Institute 2012) estimated that the average interaction worker spent close to 20 per cent of the working week looking for internal information or tracking down colleagues who could help, and that searchable institutional knowledge could cut that time by as much as 35 per cent. The human must remember and convey the relevant context from their own head. The AI must work from whatever it can be told in the moment, plus whatever its training happens to cover. The cognitive function still operates, sometimes brilliantly, but it operates on a thin substrate. Anything the organisation has previously decided, agreed, learned or written down is available only insofar as someone in the room can recall it.

With a mature knowledge base, the substrate is thick. The same conversation now draws on the considered output of every previous engagement that touched the topic. The human's own recall is supplemented rather than relied upon. The AI's general knowledge is grounded in the organisation's specific decisions and conventions. Two minds doing the work; the experience of many minds available to them.

This is what Edwin Hutchins observed when he studied a US Navy ship's bridge crew in 1995. The intelligence of the navigation was not located in any one head. It was distributed across the people, the charts, the instruments and the procedures that had accumulated over decades of practice. Take any one of those elements away and the intelligence diminishes. Take the accumulated procedures away and the crew is suddenly, urgently, much less competent than it appeared.

8.2 The knowledge base is not a document store

The single most consequential mistake organisations make about the knowledge base is to assume they already have one. They do not. They have document storage, and the two are not the same artefact.

The confusion is understandable. Both involve writing things down. Both involve folders, search and retrieval. Both, if asked, would be described as "where we keep what we know." But they are designed for different problems, and a system designed for one rarely works well for the other.

A document store is optimised for archival. Its primary user is a human being browsing folders, opening a file and reading it. The unit of value is the document. The dominant interaction is sequential reading. Length is not a problem; structure is loose; the assumption is that the reader has time to scan a table of contents and find the relevant section.

A knowledge base is optimised for cognitive access during active work. Its primary user is the cognitive function described in the previous chapter, drawing on the base as a source in fragments, often without the surrounding document being opened at all. The unit of value is the smallest piece of knowledge that makes sense in isolation: a single decision, a single convention, a single domain note. The dominant interaction is composition, pulling several small pieces together to inform the current question. Length is a problem; structure is everything; the assumption is that nothing will be read in full.

These are different design problems. They produce different artefacts. A team that treats its document store as a knowledge base ends up with long, polished, beautifully formatted files that the partnership cannot use efficiently. A team that treats its knowledge base as a document store ends up with the same long files, just in a newer system. Neither team has built organisational memory in the sense the partnership requires.

The distinction is not a matter of technology. A well-structured knowledge base can live in plain text files in a repository, or in a wiki or in a purpose-built tool. A badly structured one can live in any of those too. What matters is the shape of what is inside: short, granular, retrievable, structured for use during work rather than for reading after the fact. What Nonaka (1991) called the conversion of tacit knowledge into explicit form sits at the heart of why a knowledge base differs from a document store: the document store holds what was already explicit, while the knowledge base captures the patterns experienced practitioners hold but have rarely articulated.

8.3 What kinds of knowledge are worth keeping?

The contents of a mature knowledge base fall into a few distinct categories, each rewarding a particular shape.

There are decisions and the rationale behind them. These are best captured at the moment they are made, while the alternatives considered are still vivid and the reasons for choosing one path are still articulate. A good decision record states what was decided, why, what alternatives were weighed and what conditions might cause the decision to be revisited. It is short by design. It is also irreplaceable: a year later, no-one will remember the alternatives, and reconstructing them from scratch will cost more than the original decision did.

There are conventions: the agreed ways of doing recurring work. How the team writes a board paper. How a clinician documents a particular kind of consultation. How an engineering team handles a specific class of failure. Conventions, written down concisely, allow the partnership to apply them consistently without re-deriving them each time.

There is domain knowledge: the substantive understanding of the field as the organisation has come to hold it. Not the textbook version, which the AI partner already has. The organisation-specific version: the interpretations, the local terminology, the patterns observed across the cases this team has actually seen. Domain knowledge benefits from being captured as topical notes that can be composed together rather than as comprehensive documents that try to cover everything.

And there is institutional context: the history, the constraints, the relationships, the things that "everybody knows" until the day no-one in the room does. Institutional context is the surrounding narrative that holds the rest in place. Without it, decisions and conventions can read like arbitrary preferences rather than considered responses to a particular situation.

> *A compliance team in a mid-sized financial services firm maintains a knowledge base of regulatory interpretations. Each entry is short, a paragraph or two, and records a specific position the team has taken on a specific provision, the reasoning that led to it and any relevant correspondence with the regulator.*
>
> *The team does not maintain a master document on "our regulatory posture." That document would be both too long to use during work and too generic to capture what the team actually does. The granular entries, by contrast, are exactly what the partnership reaches for when a new question arrives that resembles an old one.*

It helps to see what entries of this kind actually look like at the granular level. The three short examples below are illustrative rather than drawn from any particular organisation, written to give a feel for the shape and length of a useful entry rather than to serve as templates.

A decision record might read:

> ***Decision: warranty cap on mid-market acquisitions, March 2025.*** *Adopted a cap of 25 per cent of consideration as the firm's standard opening position on mid-market deals between 20 and 200 million. Considered a flat 30 per cent and a tiered structure scaling with deal size. The 25 per cent figure reflects the median outcome across the firm's last forty completed transactions and aligns with where counterparty counsel has typically settled. Revisit if market practice shifts materially or if a transaction profile falls outside the 20–200 million band, where different dynamics apply.*

A convention might read:

> ***Convention: structuring a regulator response letter.*** *Open with a one-paragraph summary of the firm's position, stated plainly. Follow with the regulator's question restated in our own words to confirm the issue scope. Address each substantive point in a numbered section, citing the relevant provision and the firm's reasoning. Close with what the firm proposes to do next and by when. Avoid pre-emptive concessions and avoid hedging language that implies uncertainty the firm does not actually have.*

A domain note might read:

> ***Domain note: Regulator X engagement patterns.*** *Tends to open with a broad request and narrow as the dialogue develops, so early responses should establish framing rather than commit detail. Has historically responded well to firms that volunteer adjacent context the request did not ask for, provided it is clearly relevant. Two-week response windows are firm in writing and softer in practice, but a request for extension is better received before the deadline than after. Lead reviewer changes roughly every eighteen months; institutional memory on the regulator's side is shorter than ours.*

8.4 The Reflection phase is the engine of growth

A knowledge base does not build itself. It grows because someone, at the right moment, recognises that something has been learned and writes it down in a form that future work can use. In the Paired Cognition workflow, that moment has a name: the Reflection phase, developed at length later in the book.

Reflection is the deliberate practice of looking back at the work just completed and asking a small set of questions before closing the engagement. Were any decisions made that future decision-makers will need to understand? Did the work produce knowledge that should be added to the knowledge base for the benefit of later work? What lessons emerged that the team would want to remember next time? Where did the work expose areas of weak knowledge that should be strengthened before they become a problem?

Each of these questions, asked routinely, channels small fragments of insight into durable form. None of them takes long. Together, repeated across every engagement, they are the mechanism by which the organisation's memory accumulates rather than dissipates.

Reflection is not a separate project. It is a phase of the work itself, completed by the same partnership that did the work, while the context is still alive in everyone's mind. This is what distinguishes it from the more familiar pattern of post hoc documentation, which is almost always too late, too generic and too disconnected from the actual decisions to be useful when needed.

The discipline is small. The compounding effect is large.

8.5 Why the knowledge base compounds

The economics of a knowledge base are unusual, and the unusual economics are the strategic case for the investment.

For most organisational assets, value scales roughly linearly. Hire twice as many salespeople and revenue rises in roughly the same proportion. Buy twice as many laptops and twice as many people can do their work. The relationship between investment and return is approximately one-for-one, and there is a ceiling beyond which the marginal contribution falls.

The knowledge base does not behave this way. Each new entry is not merely additive. It combines with everything already present, creating new paths through which the partnership can reason about new questions. A single decision record is useful in isolation. A hundred decision records, well-structured and accessible, form a body of precedent that allows new decisions to be made faster and more consistently. A thousand records, layered with conventions, domain notes and institutional context, become a form of organisational intelligence that no individual could carry, and that begins to inform decisions in ways the original authors did not anticipate.

The marginal value of each addition rises rather than falls. Economists call this increasing returns, and the pattern is rare enough in operational investments that most leaders have no recent experience of it. The first year of disciplined Reflection produces a base that is useful but modest. The third year produces a base that is materially shaping the work. The fifth year produces a base that competitors cannot reproduce by hiring or licensing alone.

That last point is the strategic one. A competitor can hire similar people, license the same models and adopt the same workflow patterns. They cannot reproduce a decade of accumulated decisions, conventions and contextual understanding. In an era when frontier AI capability is broadly available to everyone, the question of where durable advantage comes from becomes acute. The knowledge base is one of the few honest answers. It is built slowly, by the organisation's own work, in a form no acquisition can shortcut.

Two underwriting teams in two insurers, both using the same model, both operating the same workflow, both staffed with comparable people. One has spent five years systematically capturing the reasoning behind unusual risk decisions through Reflection. The other has not.

The first team's partnership now reaches every decision with five years of considered precedent in active context. The second team's partnership reaches each decision largely from a standing start. The headline tooling, viewed from any reasonable distance, looks identical between the two teams, yet the output quality diverges in ways the procurement comparison never anticipated.

8.6 What changes for individuals when the organisation remembers

One of the quiet costs of intelligent work in most organisations is that critical knowledge lives in particular heads. The senior compliance officer who can recall the regulator's stance from a meeting three years ago. The architect who knows why the system was designed this way and not some other way. The relationship partner who carries fifteen years of context on a major client in their head and nowhere else. When that person is unavailable, the organisation loses access to the knowledge.

A maturing knowledge base shifts the centre of gravity. As decisions and their rationale are captured at the moment of making, as conventions are recorded as they are agreed, as domain knowledge is externalised through Reflection, the organisation begins to retain what individuals know rather than merely employing people who happen to know it.

Far from threatening the people involved, this is a release. Senior practitioners spend less time being interrupted to repeat answers they have given many times before. Their contribution moves up the value curve, towards the genuinely novel questions where their judgement matters most. The repetitive recall work, which never deserved their attention in the first place, is handled by the partnership working against the knowledge base.

The effect on the organisation is significant. Continuity of high-quality operation becomes less fragile. The risk attached to any individual moving roles, taking leave or leaving the organisation entirely is materially reduced. Employees are released from the implicit expectation of being permanently on call, because the knowledge they hold is increasingly accessible without their direct involvement.

Onboarding looks different too. In organisations without externalised institutional memory, the first months of any new role are dominated by absorption: shadowing colleagues, asking repeated questions, slowly building the mental model that their predecessors carry around in their heads. When the knowledge base is mature, the new hire arrives into a partnership that already has access to the organisation's accumulated context. They reach the point of meaningful contribution far sooner. Their early weeks are spent learning the texture of the work and the people, not reconstructing the background that should have been written down years ago.

8.7 The mistakes that hollow it out, in four shapes

Building a knowledge base is straightforward in principle and unforgiving in execution. A handful of mistakes are common enough to be worth naming.

The first is capturing too much. A knowledge base that records every meeting, every email and every passing thought becomes unsearchable noise within months. The discipline is to capture decisions and durable knowledge, not the conversational scaffolding around them.

The second mistake is the opposite, and is more common still. Teams that wait for entries to feel important enough to record almost never record anything. Important is a high bar set by hindsight. The better practice is to lower the threshold so that anything future work might plausibly need to know is written down, and to accept that pruning is easier than reconstruction.

The third is structural. A knowledge base flattened into one enormous document is functionally inaccessible: no-one will read it during work, and the partnership cannot retrieve from it efficiently. So is one fragmented across so many systems that nothing can be found. The right structure is granular enough to be retrieved by topic, consistent enough to be navigated and stored in formats that both human readers and AI partners can work with directly.

The fourth is neglecting maintenance, and a knowledge base degrades quickly if entries are never revisited. The Reflection phase handles most of this naturally, because each piece of work is an opportunity to update what it depends on. The remainder is a periodic review of older entries against current practice, refreshing what has shifted and retiring what no longer applies.

The knowledge base is a living thing. Like any living thing, it requires care to remain useful. The care is small, regular and undramatic: a paragraph captured here, an entry refreshed there, an obsolete note retired. None of it is heroic. All of it is what separates the organisations that learn from the ones that merely remember a few of their lessons until the people who learned them move on.

8.8 The investment looks small until it does not

A leader weighing where to direct effort in an AI adoption programme will see many candidates with louder claims than the knowledge base. Tooling decisions feel urgent. Pilot projects produce visible artefacts. Training programmes can be scheduled and counted. The knowledge base, by comparison, is quiet work: a discipline of capture and structure, with no launch event and no demo.

For the first year, that quietness is genuinely the experience. A team practising Reflection captures a few hundred entries. The partnership begins to reach for them. The improvement is real but incremental, and it would be easy to conclude that the investment is not paying off relative to the noisier alternatives.

Then the compounding starts. The base reaches a size at which most new questions touch material the partnership already holds. Decisions become faster and more consistent because the precedent is in active context. New hires reach productive contribution sooner because the organisation actually remembers what it knows. The same models, in the same hands, produce visibly better work, not because the models have changed but because the substrate they are working against has thickened.

The organisations that will look back on this decade with the most to show for it are not the ones that adopted the most tools or ran the most pilots. They are the ones that, while doing the work, remembered to write down what they were learning, in a form their future selves could use. That discipline is the knowledge base. It is the third node, without which the partnership stays thin and the organisation never quite gets the leverage that capable AI was supposed to bring.

With the knowledge base established as the substrate the partnership works against, the next question is how the AI enters the conversations where that knowledge is actually used.

Chapter 9

How AI Joins Your Conversations: Direct, Observed and Assisted

> *As civilized human beings, we are the inheritors, neither of an inquiry about ourselves and the world, nor of an accumulating body of information, but of a conversation, begun in the primeval forests and extended and made more articulate in the course of centuries.*
>
> — Michael Oakeshott

The dominant image of AI at work is a single one. A person at a keyboard, typing into a chat window. The image is accurate as far as it goes, and it is also incomplete. It captures one of the three forms that Paired Cognition takes in practice, and it happens to be the form that most readily fits a screenshot.

The other two forms matter more than the screenshot suggests. A great deal of professional thinking does not happen between a person and a keyboard. It happens between people, in rooms and on calls, in dialogue that has historically vanished the moment the meeting ended. Bringing AI into that dialogue, sometimes after the fact and sometimes in real time, is where the practical reach of Paired Cognition extends well beyond the chat window.

This chapter describes the three conversation types through which Paired Cognition operates inside an organisation. They are unequal in importance. Direct Conversation is familiar territory and gets briefer treatment for that reason. Observed Conversation is where the practical leverage sits today, and most of this chapter sits with it. Assisted Conversation is the future direction, and matters chiefly because it is closer than most professionals assume. The aim is to widen the working definition of partnership from one person at one keyboard to any setting in which humans externalise their thinking and an AI is positioned to make sense of what was said.

9.1 A single principle expressed in three configurations

Before the three types are introduced separately, the reason they belong together needs to be clear. Every conversation type covered here is a form of externalised cognition, the loop in which humans think out loud through speech or text and AI ingests and structures the result. Humans think out loud through speech or text. The AI ingests that externalised content and structures it into something usable. The cognitive principle does not change when the participants change. What changes is who is doing the externalising, and through which conduit the AI gains access to the material.

The three configurations differ in their participants and in the mechanics of access. Direct Conversation involves one human and an AI. Observed Conversation involves two or more humans whose dialogue is later passed to an AI for processing. Assisted Conversation involves two or more humans with an AI present during the discussion itself, listening and contributing. Each is suited to different working situations, and most professionals will move between them in the course of a single piece of work.

Recognising the shared principle is what makes the model coherent. A practitioner who has built fluency through direct chat does not need a different mental model when a meeting transcript becomes the input. The mechanism is the same. Only the conduit through which externalised content reaches the AI is different.

This is the point that most readers underestimate on a first pass. The conversation types are not three different products to evaluate or three different skills to learn. They are three expressions of the same practice, distinguished by who is in the room and when the AI processes what was said.

9.2 Direct Conversation is where most professionals begin

Direct Conversation is the form most professionals already know. A person opens a chat interface, types or dictates their thinking, and the AI returns a structured response. The exchange iterates until the artefact is fit for the practitioner's purpose. Voice-to-text input lowers the friction for senior practitioners who think faster than they type, and richer verbal interaction (where the AI speaks as well as listens) is on the way. The conduit will keep evolving, but the cognitive loop will not.

This is also where individual proficiency is built. The human is in continuous contact with the AI and can redirect or reset at any point. Iteration is fluid and mistakes are cheap. The practitioner learns what good externalisation feels like in their own voice, on their own work, before the stakes of a stakeholder meeting are added on top. Most work owned by one person, whether that is drafting advice or working through a regulatory question, will continue to be conducted predominantly in this mode. The chat window remains the foundation that the other two configurations build on.

9.3 Observed Conversation closes the gap that meetings have always created

Many of the most important conversations in professional life are not between a person and an AI. They are between people. A business analyst and a stakeholder discuss a draft requirement. An audit partner and a client work through a contentious treatment. A risk committee debates a credit exposure. A product manager and an engineer reason through a design trade-off. These conversations contain enormous cognitive value. Until recently, most of that value evaporated as soon as the meeting ended.

Observed Conversation closes that gap. The conversation is held in the normal way, typically on a video platform such as Zoom or Teams, and the transcript is exported afterwards and provided to the AI. The AI processes that transcript exactly as it would process any other externalised content. It identifies what was decided, what was discussed but unresolved, what action items emerged and what assumptions were embedded in the discussion without ever being made explicit.

A concrete example anchors the pattern.

> *A business analyst on a financial services programme is working through a draft requirements document with a senior stakeholder from the lending business. The call runs for an hour. They debate the wording of an exposure threshold and agree on a change. They surface a related question about how the new control interacts with an existing risk policy, and assign that question to the stakeholder's compliance counterpart. They also touch on an unrelated*

downstream concern that the analyst flags as out of scope for this conversation but worth recording.

After the call, the analyst exports the Teams transcript and passes it to the AI alongside the existing requirements document and a short note describing the conversation's purpose. Within minutes, four artefacts come back. The first is a redlined version of the requirements document with the agreed wording change applied. The second is a decision log entry recording the threshold debate, the rationale, and the resolution. The third is an action item assigned to the compliance counterpart, with the underlying question phrased precisely. The fourth is an entry in the open questions register noting the downstream concern flagged as out of scope, with a suggestion that it be raised in the next steering meeting.

The analyst reviews the four artefacts in roughly the time it would have taken to write up loose meeting notes. The substantive cognitive work, deciding what was said, what it meant, and what should be done next, has already been performed.

A 2025 Forrester study commissioned by Microsoft reported that meeting recaps using Teams with Copilot dropped from roughly thirty minutes to five (Forrester Consulting 2025); the figure is vendor-sponsored, but the order of magnitude tracks what practitioners report once the workflow is in routine use.

The same pattern works across professional settings. A credit committee discussion produces a structured decision memo with the dissenting views recorded as such, not flattened into a single house view. A product strategy session produces a refined option set alongside a register of the trade-offs that were considered and rejected, which is often the more useful artefact six months later when someone asks why a particular path was not taken. A client advisory call produces a record of the advice that was given and the commitments that were made on either side. In each case, the cognitive content of the conversation is preserved and converted into a working artefact rather than left to the memory of whoever happened to take notes.

9.4 Why Observed Conversation matters more than it sounds like it should

The significance of Observed Conversation lies in how much professional knowledge has historically been lost to time. Senior practitioners spend much of their working week in conversation, and the value generated in those conversations is often greater than the value generated in solo work. Steven Rogelberg's 2022 *Harvard Business Review* analysis (Rogelberg 2022) reported that managers spent on average 18 hours a week in meetings, and that 71 per cent of senior managers considered those meetings unproductive, a fair sketch of how much cognitive content evaporates today before any AI is in the room. Yet documentation has traditionally fallen to whoever was willing to take notes, with predictable consequences for accuracy, depth and the willingness of senior people to commit to the role.

When the AI processes a transcript, the documentation problem inverts. The full conversation is captured by default. The cognitive task shifts from trying to remember what was said to deciding what shape the structured output should take. The AI does not get tired in the third hour of a workshop. It has no stake in whose ideas are credited. It produces a faithful, organised rendering of what actually occurred, without the recency bias that distorts hand-written summaries written from a half-page of scrawled notes.

There is a second effect that matters as much as the first. Once the structured output of a conversation becomes routinely available, the asymmetry between people who attended and people who did not begins to dissolve. A senior leader who could not be in the room can read a high-fidelity rendering of the discussion that surfaces the substantive disagreements rather than smoothing them over. A new joiner can be brought up to speed on six months of decisions in an afternoon. The institutional memory that used to live in the heads of long-tenured staff is, for the first time, written down in a form that does not depend on someone remembering to write it down.

This has practical implications for how professionals design their meetings. A workshop run with the assumption that AI will process the transcript afterwards can be conducted differently. Participants do not need to interrupt the flow to capture every point. They can think and discuss naturally, knowing that the structuring will happen later.

The meeting becomes a thinking space rather than a documentation exercise.

That shift compounds. Once a team has run a handful of stakeholder workshops this way, the agenda design itself starts to change. Time previously spent reading minutes back to the room is reclaimed. Action items are no longer captured in real time by whoever is fastest with a keyboard; they are derived after the fact from the actual conversation. The room is freed to do the harder thing it was always supposed to do, which is think.

A practical caution is worth stating here. Observed Conversation depends on people knowing the conversation is being recorded and being comfortable that it will be processed by an AI. In some settings that is unremarkable. In others, particularly where contentious or politically charged decisions are being made, it changes the dynamics of what people are willing to say. The fix is not to record covertly. The fix is to be explicit about what is captured, who can see the structured output, and how long the underlying transcript is retained. Treated as a governance question rather than a technology question, this is straightforward to resolve. Treated as an afterthought, it produces the predictable result of a brilliant capability that nobody is willing to use on the conversations that matter.

9.5 Assisted Conversation is the natural next step, and the tooling is the gap

Assisted Conversation is the natural evolution of Observed Conversation. Instead of the AI processing the dialogue after the fact, it participates in real time. It listens as the conversation unfolds, contributes relevant information when prompted, retrieves supporting material from the knowledge base, and structures the emerging understanding as the discussion progresses. The AI becomes a third participant rather than a downstream processor.

I want to be direct about how I see this shift. Assisted Conversation is likely to be among the changes that most visibly alter the texture of senior professional work over the next several years, and it is the one most current discussions about AI in the workplace are not yet ready for. The reason is not that the technology is exotic. The reason is that the dynamics of a meeting in which the AI is a participant are different in kind from the dynamics of a meeting in which it is not. Meetings today work the way they do because the people in the room are working from incomplete and unevenly distributed memory. Take that constraint away and the basic geometry of the conversation changes. The person who has read the most papers no longer holds a structural advantage. The argument that wins is the argument that is best, not the argument made by the person who happened to remember the relevant precedent.

That is a larger change than it sounds. A great deal of senior professional authority is built on selective recall, and the rooms in which decisions are made have been organised around it for as long as anyone reading this has been working. I do not think this is something to lament. I think it is something to prepare for, because the practitioners who recognise the shift early will design their meetings to take advantage of it, and the practitioners who do not will find themselves in rooms where the rules they trained for no longer apply.

Current tooling does not yet fully support this mode, and the framing of that gap matters. Real-time transcription is mature. The orchestration required for an AI to participate usefully in a live human conversation, knowing when to speak and when to remain silent, when to retrieve and when to wait, is still under development. Although the pieces are appearing, the seamless integrated experience has not yet arrived.

This is a tooling gap, not a conceptual one. The point is worth holding onto, because it is easy to see Assisted Conversation as a fundamentally different and more advanced kind of partnership. It is not. It is the same cognitive architecture as Observed Conversation, with the processing step pulled forward into the meeting itself. Practitioners should treat it as a future direction rather than a current capability, and they should not wait for it before adopting the two configurations that already work.

When Assisted Conversation matures, the difference for participants will be tangible. A risk committee reviewing a complex exposure will be able to ask, mid-discussion, for the relevant section of the credit policy and the most recent comparable cases, and the AI will retrieve and summarise them without breaking the flow of the meeting. A strategic planning session will produce a continuously updated decision log, visible on a shared screen, that participants can challenge and amend as the discussion progresses. A requirements workshop will see new requirements appear in structured form as participants describe them, ready for review before the meeting ends.

A regional bank's investment committee is debating whether to extend a facility to a long-standing corporate client whose sector has come under pressure. Three members lean toward declining. Two argue the relationship and the client's track record warrant continued support.

As the discussion deepens, one member asks the AI for the bank's experience with similar facility decisions in adjacent sectors over the past five years. A summary appears on the shared screen within seconds: nine comparable cases, the decision in each, and the eventual outcome.

The discussion shifts from speculation to evidence. The committee still decides. The conversation it has on the way to deciding is materially better.

The value of that kind of moment is not speed. It is the quality of the thinking that becomes possible when the relevant context is in the room rather than in someone's drawer.

9.6 The continuity between Observed and Assisted is the load-bearing point

The conceptual continuity between Observed and Assisted Conversation is the most important point to grasp in this chapter. The two configurations differ only in tooling maturity, sharing the same cognitive principle: multiple humans externalising through dialogue, and an AI processing that externalised content into structured understanding. The shift from after-the-fact processing to real-time engagement does not change what the AI is doing. It changes only when the processing happens.

This matters for adoption planning. Practitioners who establish strong habits around Observed Conversation today will find Assisted Conversation an extension of what they already do, not a new skill to learn. The mental model stays the same. The tooling around it will mature without forcing the working practice itself to be rebuilt.

Organisations that wait for Assisted Conversation to be perfect before engaging with the model will miss the value available in Observed Conversation today. The transcript-and-process workflow is mature, low-cost and immediately useful. It is also the most direct route to building the muscle memory and the knowledge base that Assisted Conversation will eventually draw on.

Waiting, in this case, is the slower path dressed up as the safer one.

9.7 What this means for the design of professional work

Once the three conversation types are visible as a single cognitive architecture, the design of professional work shifts. Meetings become more deliberate. A strategic planning session is scheduled with the AI processing step built into the calendar, and the agenda is shaped for productive externalisation rather than note-taking efficiency. A client advisory call assumes that the transcript will be turned into a record of advice given and commitments made, which in turn changes how confidently the advisor speaks during the call.

Collaboration sessions can be structured to make the most of each mode. A complex problem might begin with several practitioners in Direct Conversation with their own AI partners, each developing their thinking individually and capturing it in a working note. The team then convenes for an Observed Conversation, and the combined dialogue is processed afterwards into a synthesis that integrates the individual streams. As Assisted Conversation matures, that synthesis will emerge during the meeting itself, ready for refinement before participants leave the room.

The implications run deeper than meeting design. The line between solo work and collaborative work blurs when both modes feed the same cognitive system. A piece of analysis that begins as Direct Conversation, moves through an Observed Conversation with stakeholders, and returns to Direct Conversation for refinement is not three separate activities. It is one continuous piece of thinking, distributed across people and time, held together by an AI that has been present throughout. The seven phases set out in the next chapter describe how that continuous piece of thinking unfolds inside any one of these conversation types.

That continuity is the prize. It is what makes the difference between an organisation that uses AI in its meetings and an organisation whose meetings, documents and follow-through are connected by a shared cognitive thread.

9.8 Paired Cognition is a practice, not an interface

Paired Cognition is sometimes mistaken for a particular tool or a particular interface. It is not. It is a way of working in which human thinking is externalised through conversation and AI is positioned to make sense of what was said. The interface can be a chat window, a meeting transcript or a real-time agent in the room. The principle behind each is identical, and the value compounds when an organisation treats them as expressions of the same practice rather than as separate technologies to procure and govern.

Practitioners who hold this view position themselves well for the changes the next few years will bring. Tooling will improve. Real-time participation will become routine, and conversation types not yet imagined will appear after that. The cognitive architecture beneath all of them will remain what it is now. Humans externalise their thinking. AI structures it into something the partnership can act on, and the work that emerges is work neither party could have produced alone.

The chat window was the entry point. It was never the destination.

Chapter 10

The Seven Phases of a Paired Cognition Workflow

> *Philosophers have not done justice to the distinction which is quite familiar to all of us between knowing that something is the case and knowing how to do things.*
>
> — Gilbert Ryle

A workflow that has been thought about clearly is one whose practitioners can name, at any moment, what they are doing. They can say which phase of the work they are in, which they have just left and which will come next. They know what the current phase is supposed to produce and how that output will be used. The work is legible to them while they are inside it.

This is rarer than it sounds. Most cognitive work is performed by people who can describe the deliverable they are aiming at and the meeting they are heading into, but not the cognitive shape of the hour they are in. They know what they are trying to finish. They have a much fuzzier sense of what they are trying to do right now.

The seven phases described in this chapter are an attempt to fix that.

The mechanism developed in earlier chapters, the loop in which a human externalises and an AI structures, is the engine of Paired Cognition. The phases are the operating system that runs on top of it. They give practitioners a vocabulary for the kinds of work they perform inside the partnership, and a clear allocation of responsibility within each kind. They turn an intuition about how good Paired Cognition feels into an account of how good Paired Cognition is constructed.

10.1 The phases are building blocks, not a procedure

The seven phases are Exploration, Scoping, Planning, Execution, Validation, Refinement and Reflection, and the order in which they are listed is the order in which they build on one another. Because earlier phases produce the inputs that later phases require, a team cannot Plan what it has not Scoped, and cannot Validate what it has not Executed. Each phase rests on the artefacts of those before it.

But the phases are not a checklist.

Few pieces of work use every phase. A small task may require only Planning and Execution. A research-led strategy piece may dwell in Exploration for hours before any scope is set. A routine document may collapse Scoping, Planning and Execution into a single move that takes ninety seconds and looks, from the outside, like a one-line instruction. The model is not a procedure to be followed mechanically. It is a library of named moves a practitioner can recognise, sequence, and combine.

This is the load-bearing point of the chapter. The phases are building blocks. They are not all required for every piece of work, and the discipline lies in choosing which to engage rather than in marching through all seven. A fluent practitioner moves between phases, sometimes returning to an earlier one when the work demands it, and exits the workflow when the phases the work needs have been completed.

The discipline of the model lies precisely in keeping the phases distinct. When a team conflates Exploration with Scoping, it commits to deliverables before it has understood the problem. When it conflates Validation with Refinement, it ships work that has been changed without first being checked. The phases exist as separate ideas because they answer separate questions. Treating them as separate is what gives the workflow its rigour.

What follows describes each phase in turn, with attention to what it produces and how the labour divides between the partners. The division of labour is the part of the model that most repays attention. It is also the part that most distinguishes Paired Cognition from a generic chat session in which the human and the AI take turns.

10.2 Exploration is for understanding, not for output

Exploration is open-ended collaborative reasoning. Its purpose is to build a shared mental model of a problem or domain before any decision is made about what to do.

The phase produces no deliverable. What it produces is understanding, and the understanding lives inside the partnership, a body of context that subsequent phases will draw on without ever having to reconstruct.

In Exploration the AI takes an active interrogative role. It asks questions. It structures what it hears. It synthesises across statements the human has made minutes apart, identifies the gaps in the picture they have jointly assembled, surfaces tensions the human has not yet articulated and offers framings the human can accept or reject. It is not silent. It is not merely transcribing.

The human externalises. They bring knowledge, context, intuition and constraints into the conversation, often in unstructured form, often as the questions arrive rather than in any planned order. They answer the AI's questions and raise their own. They supply the lived domain detail that no model can generate. Exploration ends when both parties recognise that the problem space is well enough understood to make decisions about it, and not before.

The trap is the impulse to produce something. A team that treats Exploration as preamble, as the throat-clearing before real work begins, will exit it too early, having accumulated a transcript but not a shared understanding. When the work is genuinely unfamiliar, Exploration is the work. The artefact it produces is the partnership's grasp of the problem.

10.3 Scoping decides what; Planning decides how

Scoping determines what will be produced. It defines boundaries, outcomes, deliverables, success criteria and explicit exclusions. Where Exploration asks what is true, Scoping asks what is in.

The AI proposes options and identifies implications. It can show the human what including or excluding a particular outcome would mean for cost, risk, and timeline, and offer two or three coherent scopes side by side with the trade-offs between them named. It does not, however, decide. Scope belongs to the person accountable for the result, and that accountability does not transfer. Priorities, what is in and what is out, the criteria for done: these are human calls. The AI's role is to make the choice well-informed, not to make the choice. A scope that has been chosen rather than drifted into is one of the strongest predictors of a workflow that will conclude cleanly. PMI's 2018 *Pulse of the Profession* benchmark (Project Management Institute 2018) recorded scope creep in 52 per cent of projects, up from 43 per cent five years earlier, and ranked it the third leading cause of project failure: a quiet vote for the discipline of choosing scope rather than drifting into it.

Planning then determines how the scoped work will be done. It produces an approach, a sequence of steps and where appropriate a set of specifications detailed enough to guide Execution. Scoping fixed the destination. Planning fixes the route.

The AI is well suited to drafting plans: ordered steps, estimated impact, dependencies, risks and specifications at whatever level of detail the work requires. The human approves the approach and makes the design decisions the AI cannot make alone: choices about method, about which trade-offs to accept, about which conventions to follow. By the end of Planning, both parties share an explicit understanding of how the work will proceed. Execution becomes a matter of running the plan rather than inventing it.

10.4 Execution is the conclusion of a long conversation

Execution is where output is produced. In a Paired Cognition workflow this is rarely a moment of cold drafting. The AI enters Execution already inside a shared mental model built through Exploration, Scoping and Planning. It knows the context. It knows the scope. It knows the approach and the standards. Producing the output is, in effect, the final step of a conversation that has been running for some time.

The AI does the bulk of the production. It writes, drafts, codes or assembles, self-validating between steps and flagging anything that does not fit the plan. Volume is no longer the constraint it once was. The constraint that mattered, alignment of output to intent, was resolved earlier.

A well-prepared Execution looks unspectacular from the outside. The AI produces. The human watches. The artefact appears.

The next chapter develops this in detail as directed execution, and it is the move that most distinguishes Paired Cognition from Task Delegation. The same AI capability, given the same eventual instruction, produces work of a different order when it has been brought into the problem first. Execution is the harvest. The crop was planted earlier. Anthropic's own 2025 engineering guidance (Anthropic 2025b) notes that "well-specified, clear, and accurate task descriptions upfront" raise model performance and reduce wasted tokens, and that recall accuracy degrades as low-signal context accumulates: the same observation, named in the model-builder's vocabulary.

10.5 Validation and Refinement answer different questions

Validation answers a single question. Does the output meet the specification it was produced against? The phase exists to separate the question of correctness from the question of preference, which belongs to Refinement. Treating them as one collapses two distinct kinds of judgement and produces work that has been changed without being checked.

In Validation the AI presents the results, walks through what it produced, and highlights any deviations from the plan. It runs functional checks, compares output against criteria, and surfaces anomalies it noticed during Execution. The human reviews against the specification, asking not whether they like the output but whether the output is what was Scoped and Planned. A failure in Validation typically returns the workflow to Planning or Scoping rather than to Refinement. Something was misunderstood, and changing the artefact in isolation will not fix the misunderstanding.

Refinement is the phase where the output changes in response to the experience of having it. Validation confirmed that the artefact met the specification. Refinement asks whether the specification, now made concrete, actually serves its purpose. It closes the gap between what was asked for and what is actually needed.

The labour reverses here. The AI applies corrections and iterates, holding the prior context steady while the surface of the artefact changes. Multiple rounds are normal and inexpensive when the underlying mental model is shared. The harder work is the human's: noticing the gaps. Reading or using the artefact in its produced form reveals things no specification could have anticipated. A sentence that reads wrong, a structure that does not flow, a clause technically correct but tonally off, a feature that works without solving the problem: the human notices these and articulates them, and the AI executes the change.

The discipline that separates Validation from Refinement matters more than it appears. A team that runs them as one phase tends to make changes it cannot account for, and to ship work whose specification has quietly shifted under the cover of editing. Keeping the two questions distinct (is this what we asked for, and is what we asked for what we needed) is what gives the partnership its honesty.

10.6 Reflection is the phase most often skipped and most consequential

Reflection is the phase most teams skip. Its absence does more damage than any of the others, because what is missing cannot be seen at the level of the individual workflow. It shows up only at the level of the organisation, over time, in the slow form of mistakes that recur and lessons that never compound. By the time the cost is visible, the engagement that could have taught the lesson is years gone, and nobody remembers what was actually learned in it.

After the work is complete, Reflection asks a different set of questions from those that drove the work itself. What was decided that should be captured? What did this work teach the organisation? What follow-up tasks have been surfaced, and who owns them? What lessons emerged about the way the work was done, separate from its content? Although these are not the questions that produced the artefact, they are the questions that decide whether the artefact taught anything.

In this phase the AI is well placed to draft: a decision record, a knowledge base entry, an internal note for stakeholders, a comparison of what was planned against what occurred, a list of tasks that arose during the work but never made it into scope. Drafting is not where the difficulty lies. The genuinely editorial work falls to the human, and it is harder, because not every decision earns a place in the knowledge base and not every observation amounts to a lesson. Most of what passes through any engagement is forgettable, and forgetting it is the right thing to do. Reflection is the act of choosing the few things that the organisation actually needs to remember, and saying clearly what they are.

When Reflection is skipped, the work still ends. The artefact still ships. The client still pays. From the inside, nothing appears to have been lost.

Yet the cycle that should have closed remains open. Decisions made in this engagement will not inform the next. The workaround discovered in Planning will not be available the next time the same obstacle appears. The novel framing developed during Exploration will live in one person's head and leave with them when they leave. This is the pattern that severs the organisational learning loop, quietly, one engagement at a time, and it is consequential enough that the next chapter is given over entirely to it. For now it is enough to name the move, place it where it belongs in the sequence, and recognise that the workflow does not end with Refinement. It ends with the deliberate act of looking back.

10.7 How phases compose in practice

A complex strategy piece may use all seven phases in sequence.

> *A consulting partner working with a client on market entry into a new geography begins in Exploration, building a shared picture of the target country, regulatory context and competitive landscape with their AI partner over the course of several sessions. Scoping fixes the question that the engagement will actually answer. Planning sets the analytical approach, the reference cases to draw on, and the structure of the final document.*
>
> *Execution produces the analysis and the deck. Validation checks coverage against the agreed scope and rigour against the agreed approach. Refinement responds to the partner's reading of the draft and to the early reactions of the client team. Reflection captures the framework, the precedents, and the follow-up actions for the engagement file, where the next partner working on a similar geography will find them.*

A simpler piece of work uses fewer phases.

> *A senior analyst drafting a routine status update for a steering committee may collapse the workflow to Planning and Execution. The human says what the update should cover and which stakeholders are reading it. The AI produces the update. There is no Exploration because the situation is well understood, no Scoping because the deliverable is standard, no Refinement because the draft is fit for purpose and no Reflection because there is nothing the organisation needs to learn from a status update.*

Different work shapes pull on different phases.

> *A bug fix in a familiar codebase typically runs through Planning, Execution and Validation, with no Exploration because the system is already understood and minimal Refinement because correctness is binary.*
>
> *A research-heavy investigation into an unfamiliar regulatory regime may live almost entirely in Exploration, producing no formal deliverable for hours but accumulating the shared understanding that makes later Scoping possible.*
>
> *A meeting note may consist of Execution and Reflection only, with the act of writing the note serving as the Reflection itself.*

The point of the model is not to insist that every workflow visit every phase. It is to give practitioners a way to talk about what they are doing while they are doing it. A team that can say which phase it is in, and which phase it is moving to next, has the operational language it needs to make Paired Cognition repeatable. It can recognise when it has skipped a phase that the work required. It can recognise when it is dwelling in a phase that the work has already left behind. It can hand a workflow to a colleague mid-engagement and have the colleague know, immediately, what kind of move comes next.

The earlier chapters of Part II established the mechanism that makes the partnership possible: externalisation, the three-node architecture, the knowledge base, the conversation types. The phases described here are the operational language a team uses to run that mechanism on real work, and they are why the next chapter can spend its time on a single distinction inside the workflow itself.

Chapter 11

Directed Execution vs. Task Delegation: Why Context Changes Everything

> *The knowledge of the circumstances of which we must make use never exists in concentrated or integrated form, but solely as the dispersed bits of incomplete and frequently contradictory knowledge which all the separate individuals possess.*
>
> — F. A. Hayek

Two practitioners use the same AI model on the same nominal task. One receives output that reads as if a thoughtful colleague had written it. The other receives a polished sequence of paragraphs that meet the brief and miss the point. Neither was using a different tool. Neither had access to a smarter model. The difference is not in the technology, and not in the prompt craft. The difference is in what the AI carried into the moment of execution.

This is the practical heart of Paired Cognition. The cleanest way to make it visible is to set the two patterns of production beside each other.

11.1 The two patterns share a surface and diverge underneath

Both directed execution and Task Delegation are legitimate ways of producing output with AI. Both have their place. The error most organisations make is treating them as interchangeable, then concluding that AI is unreliable when delegation produces shallow results on work that demanded partnership.

The two patterns differ in a single fundamental dimension. In directed execution, the AI operates inside a jointly built shared mental model that has been developed during the prior phases of work. In Task Delegation, the AI receives a discrete brief and executes against it without that broader frame. Call this second pattern the cold brief. The AI starts cold, with nothing but the instruction in front of it; whatever context exists outside the brief is invisible. Everything that distinguishes the two modes flows from this difference.

Directed execution is what happens during the Execution phase of a Paired Cognition workflow. By the time output is being produced, the AI has already participated in exploring the problem, defining its scope and planning the approach. It understands the work the way a senior collaborator would understand it.

What the AI retains in this state is substantial. It knows why the work matters and who it serves. It knows what constraints apply, including the soft constraints that emerged through earlier conversation rather than being explicitly stated. It knows how the current output fits into the larger picture. It carries the jointly built plan as an active reference, not as a buried instruction.

The cold brief works differently. The human knows what needs doing and communicates it as a discrete brief. The AI executes to that specification and returns the output. There is no ongoing mental model of the broader work, no prior phase of joint exploration, no plan that the AI helped construct. The brief is the entire context, and outside the brief is silence.

11.2 The mechanism behind the quality difference is straightforward

In directed execution, the AI self-validates its output against the jointly built plan. It checks the work it is producing against the goals, constraints and design decisions established earlier. When something feels off, the AI can flag it because it has a frame against which to recognise the misalignment. When trade-offs arise mid-execution, the AI navigates them in ways consistent with the prior reasoning rather than according to generic defaults.

Under the cold brief, the AI self-validates against the specification only. The AI can produce work that meets every requirement stated and still miss the point entirely, because the point was never stated.

From inside the AI's view, the task was completed correctly. From outside, the result feels superficial, generic or strangely off target.

This is why the same AI produces work of such different depth across different sessions. Although the model is not smarter in some sessions and dumber in others, it has more material to reason against in the ones that produce better work. Context is not decoration around the prompt. Context is what the AI thinks with.

11.3 The same legal task drafted two ways shows the gap directly

The clearest way to see the difference is to follow a single task through both patterns. The task is bounded enough to be tractable on the page and rich enough to expose the gap.

A senior associate at a commercial law firm needs a 400-word section drafted for a client advice memo. The matter concerns a contractual indemnity in an energy sector joint venture. A partner is reviewing the memo before it goes out. The client has expressed strong views about how indemnity exposure should be framed in negotiations. The counterparty's recent positions on similar clauses have been unusually aggressive.

Approached as a cold brief, the instruction to the AI might read like this:

> *Draft a 400-word section for a client advice memo on a mutual indemnity clause in an energy sector joint venture agreement. Cover the standard scope of indemnity, typical risk allocation between parties, common carve-outs (gross negligence, wilful misconduct) and a recommended position. Use a measured, advisory tone suitable for a sophisticated commercial client.*

The AI executes against the brief. The output that comes back is competent. It describes the clause correctly. It walks through the standard risk allocation. It names the usual carve-outs. It closes with a defensible recommended position written in measured prose. A reader familiar with the area would find nothing technically wrong with it.

A reader familiar with the area would also notice that it could have been written for any client in any energy sector joint venture in any jurisdiction. The associate reads the section, recognises that it is competent, and then spends the next 40 minutes rewriting it. The rewrite is not because the AI made errors. It is because the section does not yet contain the actual matter.

Approached as directed execution, the same section emerges from a different process. The associate has spent the prior hour with the AI working through the matter. The AI has heard the client's commercial priorities: protection of upstream operations, willingness to accept tighter caps on consequential loss, particular sensitivity to environmental remediation exposure given a prior incident at a nearby site. It has heard the counterparty's recent positions: pushing for broader knock-for-knock language, resisting standard carve-outs for gross negligence, citing market practice that the partner believes overstates the precedent. It has heard the partner's preferred posture in this kind of advice: close to the line on technical analysis, conservative on negotiation recommendations, plain-English on the front end of every section so the client can use the memo in their own internal discussions.

When the associate now asks for the 400-word section, the AI produces output shaped by all of this:

> *The mutual indemnity in clause 14.3 follows the broad architecture common to upstream joint ventures, but the drafting departs from market practice in two respects that warrant attention before the next negotiation round.*
>
> *First, the carve-out for gross negligence and wilful misconduct is narrower than would typically be expected in this context. The carve-out is qualified by reference to "intentional acts of senior management," a formulation which has been read narrowly in recent first-instance decisions and which would, on its face, leave the operator exposed for a wider class of supervisory failures than the client has historically accepted in comparable arrangements.*
>
> *Second, the indemnity captures consequential loss without the cap that the client has applied in its other current ventures. Given the proximity of operations to the area affected by the 2024 incident, the absence of a cap on consequential loss in respect of environmental remediation creates an exposure profile that is materially out of step with the rest of the client's portfolio.*
>
> *The recommended position is to accept the broad mutuality of the clause but to negotiate the carve-out language to reflect the standard market formulation and to introduce a consequential loss cap consistent with the client's other ventures. Both points are defensible on market practice and on the specific commercial profile of this transaction. They should be raised together rather than sequentially, given the counterparty's recent posture.*

The same 400 words. The same model. The same writing voice. The advice now reads as advice on this matter.

Notice what has changed concretely. The cold brief version names no clause;

the directed execution version identifies clause 14.3. The cold brief lists the usual carve-outs; the directed version flags a specific drafting departure ("intentional acts of senior management") and explains why it matters under recent case law. The cold brief mentions caps in the abstract; the directed version names the 2024 incident and the client's portfolio practice that puts this exposure out of step. The cold brief recommends a defensible position; the directed version sequences the negotiation move and explains the sequencing in light of the counterparty's posture.

None of these things were available to the cold brief. They were not in the instruction, and the AI had no other source to reach for. The directed execution version had access to all of them because the AI had been inside the matter for an hour before being asked to draft.

The associate no longer rewrites the section. The associate reviews it, accepts most of it, refines a sentence, and moves on. The 40 minutes saved are not the point. The point is that the AI has produced something the associate would have written, in a way that respects the partner's review standard, the client's commercial frame and the negotiation history that shaped the matter.

The pattern repeats across every domain where context drives value. A risk memo drafted from a cold brief describes generic risks. The same memo drafted within a shared model describes the risks that matter to this committee on this decision. A treatment plan summary written under Task Delegation covers the standard considerations. A summary written within a shared clinical model integrates this patient's history, preferences and prior responses. A product strategy document written from a brief describes a competent strategy. A document written within a shared model describes the strategy this team has been converging on.

11.4 Sub-delegation is how directed execution scales

The two modes are not mutually exclusive within a single workflow, and the most operationally important point about their relationship is that they compose.

Inside a directed execution workflow, the primary AI agent, the one engaged in Paired Cognition with the human, can assign discrete sub-tasks to sub-agents that operate under Task Delegation. The primary agent goes on holding the shared mental model while the sub-agents work without it, receiving a self-contained brief, executing against it, and returning the result for the primary agent to integrate.

The pattern is efficient and increasingly important as workflows become more substantial. Some execution tasks are well-bounded enough that they do not require the broader frame. Pulling a clean summary of a referenced authority. Generating a comparison table from a known dataset. Producing a cleaned transcript from a recording. These can be parallelised or run quickly without the overhead of carrying full context. The primary agent breaks them off, briefs the sub-agent, receives the output, and integrates it back into the work that remains under the shared model.

Anthropic's account of its own multi-agent research system (Anthropic 2025c) reports that an orchestrator-worker pattern (a lead agent dispatching self-contained briefs to sub-agents and integrating their results) outperformed a single-agent baseline by roughly 90 per cent on internal research evaluations, with explicit guidance that simple fact-finding warrants one sub-agent, comparisons two to four, and complex research ten or more. The shape of that guidance matters more than the headline figure: it is a sliding scale calibrated to the bounded-ness of the work, which is precisely the judgement the primary agent has to make on behalf of the partnership.

The crucial point is that the human's relationship with the primary agent does not change. The shared mental model that constitutes Paired Cognition is still being maintained between them, undisturbed by what is happening below. The sub-agents operate as Task Delegation participants inside a workflow whose overall character is directed execution. Quality is preserved at the level that matters, because the integration happens inside the shared frame.

This composition is how mature Paired Cognition workflows scale, given that not every sub-task warrants the full investment of context. The discipline lies in choosing what to delegate downward and what to retain in the partnership. Done well, throughput rises without the depth of the work suffering. Done poorly, by delegating downward the parts of the work that actually carry the strategic weight, the output reverts to cold brief quality and the primary agent reduces to a coordinator rather than a partner.

11.5 The diagnostic question is whether a competent stranger could do the work from the brief alone

Choosing between the two modes is a practical decision that depends on the work. One diagnostic question resolves most cases cleanly.

Could a competent stranger produce acceptable work from the brief alone?

If the answer is yes, Task Delegation is sufficient and probably preferable. The investment in building a shared model would not pay off because the model does not need to be rich. Drafting a standard meeting agenda, summarising a known document, formatting a list, generating boilerplate code, producing a status update from clearly stated facts: all reasonable Task Delegation candidates. A competent stranger handed the brief would produce something workable.

If the answer is no, if a competent stranger would produce something that meets the brief but misses the point, directed execution is needed. The investment in Exploration, Scoping, and Planning is what closes the gap between the literal request and the actual purpose. The output must reflect a context that cannot be reduced to a specification. Strategy work, complex legal analysis, clinical reasoning, financial advisory, complex underwriting, product architecture decisions, change-sensitive communications: all work where the nuance of the situation drives the value of the output, and where the brief alone cannot carry what the AI needs to know.

The diagnostic is humbling because it surfaces how often work that feels routine is actually context-dependent. The associate's 400-word section looked like a Task Delegation candidate from a distance. From inside the matter, it was not.

Mature practice involves moving fluently between the two modes within a single working day. A senior practitioner using AI well will sustain directed execution on the work that matters most, and delegate discrete tasks for everything that fits the cold brief profile. The choice of mode becomes a tool, used with intent rather than by default.

11.6 The early phases are the asset, not the overhead

The reason directed execution is worth the upfront investment is that the depth it produces is not available any other way. There is no prompt clever enough to substitute for a shared model that was never built. The richness either exists in the working memory between the human and the AI, or it does not. When it does, the execution that follows reflects it. When it does not, the output remains generic regardless of how the request is phrased.

This makes the early phases of a workflow into a form of capital. Time spent in Exploration, Scoping and Planning is not overhead to be minimised. It is the construction of an asset drawn against repeatedly during execution. The asset is the shared model itself, and its value compounds across every output produced inside it. Two further sections from the same matter cost the associate almost nothing. The model is already there.

The instinct to skip ahead to execution, born of a desire to see results quickly, often produces precisely the kind of shallow work that confirms the suspicion that AI cannot really handle complex tasks. The work was never the limit. The shared model the work needed was simply never built.

Once an organisation learns to recognise the difference between the two modes, the question of how to use AI changes shape. The decision is no longer whether AI is capable enough for the task. The decision is whether the task warrants the investment in shared context that would unlock that capability. The framework gives the choice a vocabulary, and the vocabulary makes it deliberate.

That capital, once built, is vulnerable: if the partnership closes the work without pausing to extract what the engagement produced, the shared model dissolves and the organisation is left with a deliverable but no lasting gain.

Chapter 12

The Reflection Phase: Why the Work Is Not Done When the Output Is

Life can only be understood backwards; but it must be lived forwards.

— Soren Kierkegaard

The deliverable is in the client's hands. The associate has hit send. The credit memo is filed, the architectural decision is logged, the report has gone out under signature. Across the partnership, the satisfaction of having finished is settling in, and the next request is already lining up in the queue. This is the moment at which the most consequential phase of the workflow is most reliably skipped.

Reflection sits in the small interval between delivery and the next piece of work, asking for 10 or 15 minutes of attention that, at the point of asking, feels like overhead. The pressure to skip it is real. Calendars are full. Curiosity about what was just produced fades the instant the work leaves the building. An hour later the texture is gone, and within a day the lessons cannot be reconstructed from memory.

The cost of skipping is invisible at the point of the decision and devastating in aggregate. Reflection is the phase that converts work into capability. Di Stefano and colleagues' Harvard Business School field experiment, conducted at a large business-process outsourcing firm, found that employees who spent the last 15 minutes of each training day writing about what they had learned outperformed peers given equivalent additional practice by 23 per cent on the final assessment, a result the team replicated in two laboratory studies (Di Stefano et al. 2016). Without that reflective pass, a Paired Cognition workflow still produces outputs, but the organisation around it stops getting smarter.

12.1 The phase that nothing punishes you for skipping

Of the seven phases introduced earlier, Exploration, Scoping, Planning, Execution, Validation, Refinement and Reflection, the seventh is dropped more reliably than any of the others. Exploration cannot be skipped without confusing the work. Scoping cannot be skipped without producing the wrong thing. Validation cannot be skipped without shipping defects. Each of those phases punishes its own absence quickly and visibly.

Reflection does not.

When Reflection is omitted, nothing immediate breaks. The deliverable has already shipped. The client has already been billed. The record has already been filed. The cost shows up later, distributed across many future engagements that will be quietly worse than they could have been. The absence is invisible to the people making the decision to skip.

The instinct to treat delivery as the end of work is older than knowledge work itself. It is the natural shape of a transaction. Something was needed, something was made, the obligation is discharged. Reflection asks the partnership to hold the door open a little longer, to look back across what was just done and ask what should be carried forward into the organisation's permanent record. It is a small ask with disproportionate consequences, and the asymmetry is exactly what makes it easy to defer.

There is a second reason it is skipped. The Refinement phase that immediately precedes it can feel like a closure of its own. The output has been adjusted in light of feedback, the partnership has agreed it is right, and the human partner experiences the satisfaction of resolution. Reflection then arrives at the moment of lowest energy and highest temptation to move on. Naming it as a separate phase, with its own discipline, is the only way to prevent it from being absorbed into the silence that follows delivery.

12.2 What the phase actually covers

The Reflection phase is a structured pass through a small set of questions, asked while the work is still fresh enough that the answers are still available.

The first concerns decisions. Were any choices made during the engagement that future decision-makers will need to understand? A decision captured at the moment of making, with its rationale and the alternatives considered, is far more useful than the same decision reconstructed three years later from partial memory and inference.

The second concerns documentation. Does anything need to be updated, revised or published as a consequence of what was just done? Internal references, knowledge base entries, external communications and process descriptions all drift when work is completed without checking what its publication implies. Reflection catches the documentation debt before it accumulates.

The third concerns the organisational memory directly. Did the work produce knowledge, conventions or domain understanding that should be added to the knowledge base? Most engagements do, but the knowledge only enters the organisation if someone notices and records it. Reflection is where that noticing is structured into the workflow rather than left to chance.

The phase also covers the softer territory that is rarely addressed elsewhere. What lessons emerged about how the work itself was done? Where did the partnership encounter weak knowledge or unclear processes? Are there follow-up tasks that need to be assigned and tracked? Who else in the organisation needs to be informed about what just happened? Each of these questions, asked routinely, channels small fragments of insight into durable form.

None of the questions are difficult. None of them require unusual judgement. What they require is the small structural commitment to ask them at all, every time, before the next engagement begins.

12.3 Why Reflection is not the same as Completion

A common error is to collapse Reflection into Completion, treating them as the same activity. They are different in kind, and conflating them quietly removes Reflection from the workflow without anyone noticing.

Completion is administrative. It closes the piece of work. Files are committed. Documents are filed in the matter management system. Statuses are updated. Time is recorded. The work is marked done in whichever system tracks such things. Completion is necessary, largely mechanical, and amenable to templates, automation and checklists. It concerns itself with the artefact.

Reflection is none of those things. It concerns itself with what the work produced, in the wider sense: the decisions made, the lessons surfaced, the knowledge that can now be added to the organisation's memory. It requires the partnership to look at what just happened and make a series of small judgement calls about what it should mean. The questions look simple but they cannot be answered by a script, because the answers depend on what actually happened during the engagement. Completion can be done well by a junior administrator. Reflection cannot.

The two phases also occupy different orientations in time. Completion looks forward, closing the work so the next thing can begin. Reflection looks backward, mining the work just finished for everything it can yield before the texture fades.

A team that does Completion well but skips Reflection has a tidy archive and a static knowledge base. The administrative trail looks healthy. Files are where they should be. Compliance is satisfied. Organisational learning quietly stalls, and no one notices because the symptoms of that stalling are diffuse and slow.

This is why Reflection deserves its own name and its own place in the workflow. Bundling it into Completion does not preserve it. The administrative work expands to fill the time, the reflective work is squeezed out, and the loss is invisible because the deliverable still went out the door. Naming Reflection separately is the small structural move that protects it from being absorbed.

12.4 What Reflection produces in practice

The case for Reflection is most easily understood by looking at what it produces inside specific kinds of work. The phase looks slightly different in each context, but the shape of the contribution is consistent: small entries, captured at the moment of clarity, that survive the engagement they came from.

> *A consulting team closing a market entry engagement uses the Reflection phase to record three things. First, the framework they ended up using to structure the competitor analysis turned out to surface insights the standard template would have missed; the partner notes the variation and the conditions under which it worked. Second, a stakeholder dynamic on the client side, a subtle disagreement between the founder and the chair on appetite for risk, shaped the recommendation in ways future engagements with this client will need to understand. Third, two of the discovery interview questions consistently produced thin answers; the engagement template should be revised.*
>
> *None of these entries took more than a few sentences. Together, across a year*

of engagements, they shape how the firm thinks about market entry.

The same phase, in a software development team closing a sprint, produces a different texture of output. Architectural choices that future engineers will need to understand are recorded with their rationale and the alternatives considered. Libraries that surprised the team in either direction are noted. Patterns of defect that suggest a gap in test coverage are surfaced rather than carried into the next sprint as latent friction. The technical decision record gains a new entry. The retrospective gains structured input rather than impressions.

> *An infrastructure team has just completed a non-trivial migration of authentication services. Reflection captures the decision to use one identity provider rather than another, the constraints that drove the choice, and the failure modes encountered during the cutover that are not yet documented anywhere. It records two follow-up items: a runbook to be written before the next on-call rotation, and a clarification needed from the security team about a residual configuration question.*
>
> *Six months later, when a similar migration begins, the team starts from these notes rather than from scratch.*

In a regulatory or compliance context, the phase produces a different kind of record again. The interpretation reached is captured along with the textual basis for it and the reasoning that connects them. The next time the same provision arises, the team starts from the prior position. Over time, the firm builds a coherent and defensible view of the regulatory landscape that survives staff changes and that demonstrates the consistency regulators look for.

> *A bank's compliance function has just closed work on the application of a new prudential standard to a particular product line. Reflection records the chain of reasoning that took the team from the rule's text to the final position, the alternatives considered and rejected, and the open questions referred upstream for legal opinion.*
>
> *When a related product line raises similar issues four months later, the analyst opening the file does not begin with the rule. They begin with the bank's prior interpretation and the reasoning that produced it.*

The pattern across these contexts is the same. Reflection turns a finished engagement into raw material for the next one, and it does so cheaply, in increments that no individual instance would justify on its own.

12.5 How skipping it severs the loop

When Reflection is consistently skipped, the costs accumulate in patterns that are recognisable to anyone who has worked in a knowledge-intensive organisation for long enough.

Decisions are remade because no one can find the rationale for the original choice. Edge cases are rediscovered because the previous encounter was never recorded. The same workshop is run a second time, because the lessons of the first were not written down. Senior staff are interrupted to answer questions that the organisation has, in fact, already answered, but only inside a head that has since moved on.

The knowledge base, which depends on Reflection for its growth, begins to lag the actual practice of the organisation. New entries trickle in only when someone has time for a documentation push, which is to say rarely. The base ages, becomes inconsistent with how work is now done, and gradually loses the trust of the partnerships that should be drawing on it. A trusted knowledge base reinforces its own use. A neglected one teaches people to ignore it.

Subtler than any of these, without Reflection the partnership itself does not learn. The human partner does not notice the patterns that would make the next engagement go more smoothly. The cognitive function cannot be guided towards the conventions and emphasis that future work would benefit from, because no one has surfaced those conventions explicitly. The same friction recurs because no one paused to name it.

Across many engagements, this absence severs the feedback loop that makes Paired Cognition compound. The three-node architecture developed earlier, in which human, AI, and knowledge base together carry the cognitive function, depends on the third node growing in step with the work being done. Reflection is the mechanism by which that growth happens. Without it, the knowledge base is whatever it was when the organisation last made a deliberate effort to build it, while the work continues to move on.

Each piece of work then becomes a discrete event rather than a contribution to organisational capability. The model still works, but only at the level of individual deliverables. Its larger promise, that every well-handled engagement leaves the organisation slightly smarter, never materialises.

12.6 How Reflection compounds when protected

The opposite case is what makes Reflection worth defending. When a partnership routinely closes its work with a structured reflective pass, the knowledge base grows steadily as a by-product of getting things done. Decisions enter the record. Conventions are clarified. Domain knowledge is externalised in small, useful increments. Workflow improvements suggest themselves as patterns become visible across many engagements. The After-Action Review, formalised by the United States Army following the 1973 Yom Kippur War and operational at the National Training Center from 1981, is widely cited as one of the most successful organisational learning methods yet devised, distinguished by rank irrelevance and a forward-looking orientation that feeds each lesson directly into the next iteration (Cates 2022).

Over months and years, this accumulation transforms the organisation's relationship with its own work. Reliance on individual memory diminishes because the memory now lives somewhere accessible. New team members arrive into a partnership that already holds the organisation's accumulated context, rather than having to absorb it through years of proximity to senior staff. Continuity through transitions becomes a property of the system rather than a matter of luck.

The compounding effect is not abstract. A team that captures one or two decisions per engagement, across hundreds of engagements, eventually holds a body of precedent that allows new decisions to be made faster and more consistently. A practice that records weak-knowledge observations across many projects eventually identifies the systemic gaps that no single project could surface. A function that notes process friction in passing, every time it occurs, eventually has the data to redesign the process.

This is why Reflection is properly described as the engine of organisational learning rather than as a tidy professional habit. It is the structural feature of the workflow that converts each instance of work into a contribution to durable capability. Without it, an organisation can perform brilliantly on a per-project basis and still be no smarter at the end of the year than at the beginning of it.

12.7 Making Reflection a habit the organisation can rely on

The shift from intention to habit is the practical work of installing Reflection in an organisation. It does not happen by exhortation. It happens through small structural moves that make the phase difficult to skip and easy to perform.

The first move is structural. Reflection should appear as an explicit phase in workflow templates, with its own checklist and a small block of dedicated time at the close of each engagement. If it is not on the schedule it does not happen. If it is on the schedule but bundled with Completion, it does not happen. The phase needs its own line.

The second move is to lower the threshold. Teams that wait for entries to feel important enough to record almost never record anything. The discipline is the opposite. Capture small things willingly, on the assumption that pruning later is easier than reconstructing. A two-line decision note is a complete and useful contribution. Perfection is the enemy of accumulation.

The third move is to make Reflection visible. Teams that share what they captured in their last Reflection, even informally, in a weekly stand-up, a monthly knowledge review or a brief written note circulated to the practice, normalise the discipline and demonstrate its value. The knowledge base entries that emerged from last month's engagements become reference points in this month's work. The connection between the discipline and the payoff becomes legible to everyone involved.

The fourth move is to remember what is at stake. Although Reflection is easily described as a professional nicety or an additional cost of working with AI, it is the phase at which the partnership of human, AI, and knowledge base closes its loop and begins to compound. Without it, each engagement is an isolated event. With it, the organisation becomes one of the small number that genuinely accumulate intelligence over time rather than spending years performing work and arriving back where they started.

12.8 The close of Part II

Part II has worked through the interior of Paired Cognition. Externalisation gives it its working substrate. The three-node architecture distributes its cognition across human, AI, and knowledge base. The knowledge base gives the partnership a memory that lasts. The three conversation types describe how the partnership joins human dialogue. The seven phases give a workflow its shape, and the distinction between directed execution and Task Delegation explains why the same model produces such different work in different hands.

Reflection is the closing piece because it is the phase that protects the others from being done in vain. Everything established across these chapters depends on it. The work is not done when the output is. It is done when the organisation has been changed, however slightly, by the experience of doing it.

Part III turns from the interior of the model to the question of how to install it. Strategy, workflow mapping, automation, training, embedded systems, pilots: the operational disciplines through which the framework becomes the way an organisation runs. The next chapter begins where every adoption effort should: with a strategy that names commitments rather than aspirations, and that knows the difference.

Part III

Making It Happen

The implementation path from strategy and readiness assessment through to embedded adoption: workflow mapping, automation, training, embedding and piloting.

CHAPTER 13

AI Strategy: From Readiness Assessment to Actionable Roadmap

> *Bad strategy is long on goals and short on policy or action. It assumes that goals are all you need.*
>
> — Richard Rumelt

Most documents that organisations call an AI strategy are not strategies but positioning documents. They name pillars, sketch a capability map, narrate a vision and circulate through a steering committee that approves them with the relief of a body that has produced something tangible. Six months later they sit unread, having shaped no decision that would have been made differently in their absence.

A strategy is a different artefact entirely. A strategy is a set of binding commitments: where an organisation will apply AI, how it will be applied, what has to change to support it and in what order. Commitments can be met or missed. Vision statements cannot.

The gap between these two artefacts is the practical subject of this chapter, and it is the place where most AI initiatives quietly fail before any tool is procured or any workflow is changed. It is also the right place to begin Part III. The chapters that follow take up the operational disciplines through which a framework becomes the way an organisation actually runs. Strategy is the first of them because nothing downstream can be coherent until the commitments above it are.

13.1 What does an AI strategy actually decide?

MIT Sloan Management Review and BCG's 2025 ninth global AI study (MIT Sloan Management Review and Boston Consulting Group 2025), surveying 2,102 respondents across 116 countries, found that 47 per cent of organisations had no strategy for what they were going to do with AI, even as 35 per cent had begun deploying agentic systems and another 44 per cent planned to. The absence is not for want of activity; it is the activity running ahead of the commitments that would make it coherent.

A working AI strategy answers four questions with specificity, and a document that does not answer them is doing some other job.

The first question is *where*. Which workflows are in scope, named at a level of detail that would let a workflow owner recognise their own work in the description. "AI in claims" is not an answer. "AI in complex bodily injury claim assessment, in the analysis phase between intake and reserve setting" is.

The second question is *which model*. Each scoped workflow has to be assigned an operating model, Paired Cognition, Task Delegation, Autonomous Agency or Embedded Intelligence, because the model determines almost every subsequent design decision. The same claims function can require all four, applied to different sub-workflows, and a strategy that does not name them at this level will design the wrong tooling for at least some of them.

The third question is *what changes*. Because AI does not slot into existing workflows untouched, it changes how decisions are sequenced, who needs to see what, where the audit trail is created, what skills the team requires and how the work is governed. A strategy that does not name these changes will be surprised by them later.

The fourth question is *in what sequence*. Some moves depend on others, and some require capability the organisation does not yet have. The order in which moves happen determines whether the organisation builds momentum or stalls, and a strategy that lists moves without sequencing them has not finished its work.

The point of this framing is not procedural rigour for its own sake. It is that strategies and positioning documents fail in different ways, and the second kind of failure is more useful than the first. A positioning document fails when its language goes stale; nobody can tell exactly when, and nobody is held to anything. A strategy fails when a commitment is missed, and the missed commitment surfaces a real constraint the organisation now has to address. The second kind of failure is the engine of learning. The first kind is the absence of one.

13.2 Strategy is closer to architecture than to vision

A useful mental model for AI strategy is the architecture decision record rather than the transformation programme. An architecture decision record names a problem, the options considered, the choice made, the reasoning behind it and the conditions under which it would be revisited. It is binding until something changes that would warrant revisiting it. It is also, deliberately, short.

Most AI strategies would benefit from being shorter. The temptation toward length is the temptation to substitute volume for decisions. A 30-page document with extensive vision narrative, vendor comparisons and risk taxonomies often contains fewer real decisions than a six-page document that names workflows, models, gaps and dates.

> *A regional insurer produced a 60-page AI strategy in early 2025. It contained five strategic pillars, a maturity model, an organisational capability map and a portfolio of 14 possible use cases. It named no specific workflow, no operating model, no readiness gap and no date. Eighteen months later the document had been quietly replaced with a different 60-page document by a different consulting team. The same workflows remained un-redesigned. The same gaps remained unaddressed. The first document had been mistaken for progress, which is the most expensive kind of error in this work.*

The corrective is to treat the decisions as the deliverable, not the document. The document is the artefact in which the decisions are recorded. When the decisions are clear, the document becomes shorter and more useful in the same motion.

13.3 Six dimensions of readiness, not one

Readiness is the bridge between aspiration and commitment. It is also the dimension most often skipped. Organisations want to know what they will do with AI. They are less eager to know what they currently cannot do, because the answer almost always implies investment they had not budgeted for.

The temptation in readiness work is to collapse it into a single score, a maturity level, a colour on a heatmap. The shape of the question resists this. Readiness is the conjunction of six conditions, and any one of them, if absent, will block adoption regardless of how strong the others are. Averaging across the six produces a number that hides the only thing worth knowing.

The first dimension, *workflow readiness*, asks whether the work itself is understood in enough detail to redesign it. This is harder than it sounds. Most organisations believe they know how their workflows run, and discover, on examination, that what they know is an idealised version that no one actually follows. The documented process and the lived process diverge most at the points where AI would have the greatest impact: the judgement-heavy moments, the escalations, the workarounds that absorb edge cases.

> *A mid-tier law firm asked a partner to describe how a corporate transaction memo gets drafted. The partner produced a clean four-step account: associate research, draft, partner review, client copy.*
>
> *Two days of observation produced a different account: 13 steps, four parallel research threads, two informal escalations to a tax specialist who was not on the matter file, and a rolling dialogue between the associate and the partner that shaped the draft long before it was formally reviewed. The four-step version would have produced a workflow design that automated the wrong things and missed the parts where AI could actually contribute.*

The gap between described and observed practice is itself a readiness signal. An organisation that cannot describe its own workflows accurately is not blocked from adoption, but it is blocked from designing for adoption, which amounts to the same thing.

Data readiness is the second dimension, and the one most often misdiagnosed. The hard cases are rarely about data quality in the abstract. They are about plumbing. A bank may have excellent credit data inside the originating system and excellent customer interaction data inside the CRM, with no path between them that an AI tool could traverse without manual export. The data itself is fine, and what fails is the plumbing between the systems that hold it. The strategic question is which subsets are reachable in time for the moves the strategy proposes, and which require infrastructure work the strategy itself has to commission.

The third dimension, *team readiness*, covers the people who will work with the AI day to day: their existing skills, their willingness to change practice and the local management capacity to support the change. A team that has been told for two years that AI will replace them is not ready, even if every other dimension is in order. The pilot will quietly fail. People will use the tool just enough to satisfy reporting and not enough to find out what it can do. A team that has experimented with consumer AI tools on its own initiative and seen genuine value is often readier than a team that has been formally trained but never used anything in a real workflow. Posture is harder to repair than skill.

Governance readiness, the fourth dimension, asks whether the organisation has the decision-making structures to oversee AI use. This is most acute in regulated environments. A bank cannot deploy a model that influences lending decisions without a clear line of accountability, a model risk framework, a monitoring approach and a rollback plan. Governance readiness is not the same as having a policy document. It is having the people, the meetings and the authority to make decisions when something needs to change. Where it is missing, the failure mode is specific: a pilot reaches the point of production deployment and stalls, because no one has the authority to say yes and no one is willing to be the person who said yes when things go wrong.

The fifth dimension, *cultural readiness*, is the softest and most often the decisive one. It is also the dimension organisations are least willing to score honestly, because the assessment cuts close to questions of leadership and identity. The relevant elements are the organisation's tolerance for iteration, its appetite for transparency about failures and the working relationship between its technical and business functions.

A culture that punishes near-misses will not surface the issues that need to be surfaced. A model will drift, a workflow will misfire, a handler will notice something off and the noticing will not travel. By the time the issue reaches anyone empowered to act, it has been a problem for months. A culture in which the technology team and the business operate as separate kingdoms will struggle with Paired Cognition specifically, because Paired Cognition requires the two to design workflows together rather than throw requirements over a wall.

Culture is not changed by a strategy document. But the strategy has to know what culture it is operating inside, because every other readiness investment lands in that culture and either takes root or does not. Of the six dimensions this one is least amenable to a checklist and most likely to determine the fate of the work.

Technical readiness, the last of the six, is the infrastructure dimension: identity, access controls, integration capacity, the ability to run or call models in a way that meets the organisation's security posture. For a mid-sized professional services firm this often resolves to the limits of the current Microsoft or Google environment; for a hospital network it becomes a much harder conversation about clinical systems and data egress. Technical readiness is the dimension most easily mistaken for the whole of the question, particularly inside IT functions, and is almost never the binding constraint. Naming it last is deliberate. A strategy that begins with infrastructure has usually let the IT function set the agenda for an organisational problem.

The six dimensions form a system, and the system has a single rule: the lowest score binds. An organisation with every infrastructure capability and no workflow ownership cannot deploy. An organisation with mature governance and a culture that punishes near-misses cannot deploy either, even if the technology works on the day of launch. In practice the binding constraint is usually cultural or workflow-related rather than technical, which is part of why technically-led strategies so often miscarry. BCG's 2025 *Build for the Future* survey (Boston Consulting Group 2025b), drawing on data from roughly 1,250 executives, restated the people-process-versus-algorithms split: AI leaders allocate around 70 per cent of their AI investment to people and process change, 20 per cent to data and technology, and only 10 per cent to algorithms. Most failed AI programmes invert this ratio. The purpose of looking at all six dimensions is to find the binding constraint before it surprises the programme, and to design the strategy around closing it rather than around the dimensions where the organisation already feels capable.

13.4 Conducting the assessment without flattering the organisation

A meaningful readiness assessment does not begin with a survey. Surveys produce confident answers that are usually wrong, because respondents describe the organisation they think they should be rather than the one they are. The questions sound neutral; the answers are aspirational. The aggregate is a portrait of the organisation as imagined by people who would prefer it to be in better shape.

The assessment begins instead with workflow observation and document review, with interviews layered on top.

An engineering organisation wanted to assess readiness for AI assistance in its site reliability function. The temptation was to begin with an interview round. Instead the assessment started with a quarter of incident post-mortems read end to end, alongside a sample of on-call handover notes.

By the time the interviews began, the assessor knew which incident classes were under-investigated, which runbooks were stale and which engineers were carrying disproportionate institutional knowledge in their heads. The interviews became conversations about specific situations the engineers recognised, not generic questions about practice. The quality of the assessment was determined before anyone was interviewed.

The questions worth asking are concrete. Where do people already get stuck. Which decisions take disproportionately long relative to their importance. Where does information get rekeyed. Which judgements are routinely second-guessed. Which workflows have a long tail of edge cases that consume disproportionate effort. These are the seams along which AI is most likely to add value, and they are also the seams that reveal which operating model fits.

A useful assessor leaves an organisation able to say something it could not say before: not "we are 60 per cent ready" but "we are blocked on three specific things, in this order, and the rest is workable." That kind of finding produces decisions. The maturity-score finding does not.

13.5 The three gaps that recur

Most organisations entering this work hit one or more of three specific gaps. They are recurrent enough to anticipate, and naming them in advance shortens the assessment.

The first is the absence of workflow ownership. There is often no single person accountable for how a process actually runs end to end, only a series of functional managers whose authority stops at the handoff points between their teams. AI redesign requires a workflow owner: someone with the authority to change how the work flows, not merely how their part of it is performed. Where this role does not exist, it has to be created before adoption, not after, and the strategy has to name who will hold it.

The second is data plumbing rather than data quality. Most organisations have data of acceptable quality somewhere. They cannot get it to where it would be useful without significant integration work that nobody has been funded to do. A wealth manager might have excellent client information in the CRM and excellent portfolio data in the trading system, with no automated path between them. The strategic implication is that data work and AI work are usually the same work, and budgeting for one without the other is a category error.

The third is governance maturity. In regulated industries this is usually the binding constraint on the speed of adoption. The model risk function in a bank, the clinical governance committee in a hospital, the compliance team in an insurer: these groups are not obstacles to be routed around. They are stakeholders whose involvement determines whether a deployment will survive contact with the regulator. Strategies that ignore them produce pilots that never go to production.

All three gaps share a common shape: they are organisational, not technical. They cannot be closed by procurement. They are closed by decisions about authority, by investment in integration and by the deliberate maturation of governance structures. A strategy that names them honestly is already partway through closing them.

13.6 Operating model selection is part of strategy, not a downstream choice

A strategy that does not specify which operating models apply to which workflows is incomplete in a load-bearing way. This is the point at which most strategy work falls short, and the failure is often invisible because the document still reads as substantive.

The four models are not interchangeable. They imply different workflow designs, different governance structures, different skill requirements and different success measures. A workflow designed for Paired Cognition and then implemented as Task Delegation will systematically under-produce; a workflow designed for Task Delegation and run as Paired Cognition will systematically waste senior time. The mismatch is not a tooling problem and cannot be fixed by a different vendor.

An insurer working through its strategy named four AI initiatives in claims. A claims handler working complex bodily injury matters was identified as a

Paired Cognition workflow, where the AI surfaces precedent and pattern across comparable matters and the handler integrates them with judgement and the file in front of her. Claims intake, straightforward submissions routed and acknowledged, was identified as Embedded Intelligence work, configured at design time and invoked without runtime human attention.

Subrogation document summarisation, high volume and well specified, was assigned to Task Delegation. Fraud screening, where the model operates inside tightly defined limits and escalates uncertain cases, was assigned to Autonomous Agency. One function. Four operating models. Four distinct designs, each with different governance, different success measures and different team implications.

A strategy that performs this assignment workflow by workflow has done the load-bearing work. A strategy that stays at the level of "we will use AI in claims" has not. The first will produce coherent investments and tractable governance. The second will produce a portfolio of disconnected projects and the slow recognition that nothing fits together.

13.7 From assessment to roadmap

A roadmap is the translation of readiness findings into a sequence of moves. The sequencing logic is straightforward in principle: do the things first where readiness is highest and value is clearest, use those wins to build capability and credibility and use the proceeds to address the gaps that block higher-value work later.

In practice this means most organisations should not begin with their most strategic ambition. A consultancy whose ambition is AI-augmented advisory work should probably not start there. It should start with research synthesis, document drafting or proposal preparation, where the workflows are well understood, the data is contained and the operating model is naturally Task Delegation. The advisory work, which calls for Paired Cognition, comes later, after the firm has learned how it actually works with AI.

Each move on the roadmap should specify the workflow being changed, the operating model being applied, the readiness gaps that have to be closed first, the success measures and the decision point at which the move is judged a success or a failure. A roadmap without these specifics is a wish list dressed in calendar quarters.

The decision points matter as much as the success measures. A roadmap that names a date on which the question "did this work" will be answered, and names who will answer it, behaves differently from a roadmap that allows initiatives to drift into permanent pilot status. Pilots that cannot fail also cannot succeed, because nothing forces the organisation to act on the result.

13.8 The discipline of avoiding transformation theatre

The label for what a strategy fails into when its commitments are not real is transformation theatre: the production of visible activity around AI without the underlying operational design that would make AI valuable. Workshops happen. Frameworks circulate. Executives become fluent in the vocabulary. Meanwhile, in the workflows where decisions are actually made, the relationship between humans and AI remains undefined.

The missing element is operational design at the level of specific workflows.

A strategy document avoids theatre when it makes commitments that can be falsified: workflows named at a level of detail their owners would recognise, operating models assigned to each, readiness gaps that have to close before each move can begin, dates that someone is willing to be held to, owners who would be present in the room when the question is asked. If the document could survive being read aloud at a steering committee in 12 months as a list of things that did or did not happen, with named owners present in the room, it is a strategy. If it could not, it is theatre.

The most common strategy mistakes follow from treating the document as the deliverable rather than the decisions as the deliverable. Strategies that are too broad, that focus on tools rather than workflows, that omit governance design or that skip the workflow analysis entirely, all share a common origin: the document was the goal. When the decisions are the goal, the document becomes shorter, more specific, and considerably more useful, and the work that follows it has somewhere to go.

The remaining chapters of this part take up that work. Workflow mapping is the discipline that makes the strategy concrete. Automation, training, embedding and piloting are the operational moves through which the strategy becomes the way the organisation actually runs. The strategy is the first commitment.

Everything that follows is whether the commitment is kept.

Chapter 14

Workflow Mapping: The Foundation of Intelligent AI Integration

> *The formal scheme was parasitic on informal processes that, alone, it could not create or maintain.*
>
> —James C. Scott

AI cannot be designed into a workflow that has not been understood, and although the sentence sounds obvious, the discipline it demands is the one most often skipped. The rush to design, to pick a tool, name a use case, brief a vendor, runs ahead of the slower, less glamorous work of seeing how the work actually happens.

What replaces seeing is assumption. Documented procedures stand in for current practice. Org charts stand in for information flows. The configuration of last decade's software stands in for the present-day operating reality. Each is a partial truth. Partial truths are exactly the kind of input that produces an AI design which looks coherent on paper and breaks the moment a real case touches it.

The previous chapter argued that an AI strategy is incomplete without operating models assigned to specific workflows. This chapter takes up the prior question. Before a workflow can be assigned a model, it has to be understood, and understanding it is its own discipline. The work of this chapter is to make the case for mapping as a separate, prior practice, and then to draw a careful line between two kinds of mapping that look similar from a distance and are not the same.

14.1 Mapping is the work of admitting what is not known

Every serious engagement begins with the same admission. Nobody fully understands how the work actually happens.

Senior leaders describe the workflow at the level they are responsible for: targets, exceptions, escalations. Frontline practitioners describe it at the level they live in: their own tasks, their own irritations, the parts they have personally optimised. Process documentation describes the version that existed when the documentation was written, sometimes years prior. Software systems describe what was automated at some earlier point, not what humans now do around the gaps in those systems.

None of these views is wrong. None is sufficient. The real workflow lives in the seams between them.

Designing AI into a workflow that has not been mapped produces a recognisable pattern of failure. The model is asked to perform a step nobody can describe with precision. Outputs land with someone who was never the intended recipient. Decisions that looked routine turn out to depend on a piece of context that lives only in someone's memory. The pilot underperforms. When the steering committee asks for better prompts, better prompts will not help, because the underlying problem, that the workflow was never properly understood, remains untouched.

Mapping precedes design because design requires a target. Without a current-state model, every choice about where AI should sit, what it should do, and who it should serve becomes guesswork dressed up as strategy. Mapping is not a preliminary courtesy, but the work that produces the artefact everything downstream depends on. Skip it and the rest is theatre.

Thoughtful engagements treat the existing workflow as the primary source of truth. The team or system as it operates today contains the constraints, dependencies, and tacit logic any new design must accommodate. AI is then designed into that reality at practical detail, with the goal of working solutions rather than aspirational diagrams.

14.2 What does a workflow map capture, beyond a sequence of steps?

The deliverable is not a flowchart. A flowchart is one of its outputs, and not the most important one.

A workflow map records the actors involved, both human and software, and the roles they play. It identifies the discrete tasks performed, the decisions made and the criteria, written or otherwise, that govern those decisions. It tracks the handoffs where work transfers between parties.

It captures information flows. What inputs does each step require, where do they originate and in what form do they arrive? What outputs are produced, who receives them and what happens if they are late or wrong? These flows expose the most fragile parts of a workflow, where small disruptions cascade into large delays.

Tools and systems form another layer. The map records which applications, repositories, spreadsheets and channels are used at each step. In professional services this layer is often chaotic. A single matter, claim or engagement may pass through a practice management system, a shared drive, three email threads and a personal notebook before it is resolved. The map names this honestly rather than tidying it.

Beyond visible structure sit the elements that make workflows actually function. Implicit knowledge is the experience and judgement practitioners apply without articulating. Bottlenecks are points where work piles up, often invisible from outside the team. Exception handling describes what happens when the standard path fails, which, in many workflows, is most of the time.

A map that captures only the visible structure is a brochure. A map that captures the implicit layer is a tool.

14.3 Mapping for automation and mapping for partnership are not the same exercise

This is the distinction the rest of the chapter rests on. Two organisations can map the same workflow and produce two entirely different artefacts, both correct, neither sufficient as a substitute for the other. The difference is what the map is being put to.

Mapping for automation, where software is intended to replace a human step entirely, is a discipline of specification. Inputs, outputs, decision rules and edge cases must be captured with enough precision that a deterministic system can execute them. Anything that cannot be specified cannot be automated. The discipline of an automation map is to keep pressing on every step until it is either fully described or revealed as containing judgement that resists description. The map's value is measured in how completely it bottoms out.

Mapping for partnership, where AI is intended to think alongside a human rather than replace them, is a discipline of cognitive shape. The map must surface how the practitioner reasons. What questions do they ask themselves at each decision point? What context do they hold in their head while making it? Where do they draw on experience that resists articulation? Which moves feel routine, and which require them to slow down and look up? The map's value is measured in how faithfully it renders the texture of the work.

These two orientations produce different artefacts from the same underlying workflow. Consider a contract review process inside a corporate legal team.

> *An automation map of the review process catalogues clause types, flag conditions, routing rules and metadata to be extracted. It captures which clauses must always be reviewed, which can be passed through if they fall within standard parameters and what triggers escalation to a senior lawyer. The map is dense with thresholds and rules. Its purpose is to be detailed enough that a system can execute the routine portion of the review without further human input.*
>
> *A partnership map of the same process catalogues the patterns of attention a senior lawyer applies. Which provisions does she read closely on a first pass and which does she scan? When she sees an unusual indemnity, what does she compare it against in her head? What past matters does the language remind her of? Which combinations of clauses change the meaning of an individual clause that would be unremarkable on its own? The map is dense with reasoning and reference. Its purpose is to give an AI partner enough of the lawyer's working context to think alongside her on the parts of the work that are not routine.*

Both maps are useful. They are not interchangeable, and confusing them is the most expensive error available at this stage. An organisation that uses an automation map to design a Paired Cognition workflow will produce a system that handles the easy cases, which the lawyer was already happy to handle, and goes silent on the cases where partnership would actually create value. An organisation that uses a partnership map to design an Embedded Intelligence pipeline will produce a system that has rich context but no clean specification, and will generate inconsistent outputs at scale. Both versions of this error are expensive. Both are common.

The purpose of the map shapes what it must reveal. A team that knows which kind of map it is making is already most of the way to a usable one.

14.4 What mapping surfaces is rarely what the organisation expected

Workflow mapping produces findings that surprise the organisations being mapped, and the surprises are almost always about the organisation's own understanding of itself rather than about the world outside it.

The pattern usually starts with location. Leaders typically assume the highest-value AI integration sits at the start of a process, where the most cases pass through. Mapping frequently shows the opposite. The leverage lies further in, at the points where senior judgement is applied to fewer, more consequential calls. Volume sits at the front. Value sits in the middle. A strategy that follows the volume designs AI for the cheap part of the work. A strategy that follows the value designs it for the part where being right matters.

Once attention moves to the middle of the process, a second observation tends to follow. The middle is where the bottlenecks live, and most of them are invisible from outside the team.

> *A claims process appears, on the dashboard, to run on a five-day cycle. Mapping reveals that three of those days are spent waiting for a single specialist to review a particular kind of file. Although the dashboard shows throughput, it does not show the specialist's queue.*
>
> *Until the queue is named, every conversation about cycle time circles around the wrong levers, adding capacity at intake, refining the triage rules, replatforming the case management system, none of which touch the actual constraint.*

This is part of a broader documented pattern. Deloitte's 2025 *Global Process Mining Survey* (Deloitte 2025a) found that organisations applying systematic process mining routinely surface manual rework that is invisible to dashboards and standard operating procedures: work the official process record does not capture but the event log makes legible. The dashboard reports the workflow management has agreed to recognise. The map, like the event log, reports the workflow that is actually being run.

A diligence workflow that looks parallel turns out to serialise on one partner's review. A loan origination workflow that looks straight-through turns out to bounce three times between the credit team and the relationship manager because of a documentation step that nobody owns. These bottlenecks are obvious to the people standing in them. They are invisible from the outside.

Naming the bottleneck usually leads to the next question: why does the work pile up there? The answer is almost always the same. The constraint is a person, and the reason it is a person is implicit knowledge.

> *A senior underwriter knows which broker submissions are reliable and which need extra scrutiny. He cannot tell you the rule he is applying, because there is no rule, only a long history of submissions, a feel for the brokers, a sense of which markets are running hot and a memory of the last three losses that came in dressed as routine business.*
>
> *None of this knowledge appears in any document. None of it is captured in any system. When he retires in 18 months, none of it leaves with a handover.*

A tax partner knows which authority will accept a particular position. A relationship manager knows which clients tolerate a slow response and which will move banks if a call goes unreturned. A clinical lead knows which referring GPs send patients who are sicker than the referral letter suggests. This knowledge is the workflow, in a meaningful sense. Its loss is rarely planned for, and a Paired Cognition design depends on it being articulated before the practitioner who holds it leaves.

Once the knowledge is named, a wider drift comes into view.

By the time the implicit knowledge is on the table, the relationship between what the organisation believes about its own workflow and what is actually happening has usually broken down. The official process for onboarding a client may be a six-step checklist. The real process involves 17 informal steps, three workarounds for system limitations and a recurring side conversation with the legal team that nobody has documented. The official process is what gets shown to auditors. The real process is what produces the work. AI designed to the official process automates a fiction.

These findings arrive in roughly that order, and they reinforce each other. The shift in attention from volume to value points the team at the middle of the workflow. The middle is where the hidden bottlenecks live. The bottlenecks are usually a person, and the reason is knowledge that has never been written down. The undocumented knowledge is part of a wider drift between the workflow on paper and the workflow in practice. A mapping engagement that does not arrive somewhere along that chain has not gone deep enough.

14.5 A mapped workflow makes the operating model choice almost mechanical

Once the workflow is mapped, the choice of operating model at each integration point is no longer a matter of preference. The structure of the work tells the designer what fits.

A step characterised by ambiguity, contextual judgement and frequent exceptions points toward Paired Cognition. A step characterised by clear specifications and well-bounded inputs points toward Task Delegation. A step that recurs in similar form, with stable inputs and well-understood success criteria, points toward Autonomous Agency, provided the guardrails can be specified. A step that is part of a larger pipeline, with no genuine human decision at execution, points toward Embedded Intelligence.

The four models stop being independent choices and start being a portfolio. A single end-to-end workflow may include all four.

> *A healthcare administration team handling provider claims has, on inspection, four distinct kinds of work happening inside what is described as one process. Intake classification, sorting incoming claims by type, urgency and routing destination, is high-volume, well-specified and tolerates no variability. That is Embedded Intelligence work. Eligibility checks against policy and benefit rules are repeatable and bounded but require some interpretive latitude in edge cases. That is Autonomous Agency, with the edge cases escalated. Drafted responses to specific provider queries, where the question is clear and the answer template is established, are Task Delegation. Complex appeals reasoning, where a clinician and the AI work through whether a denial should be reconsidered against the medical record and the policy text, is Paired Cognition.*
>
> *Designed as a single workflow with one operating model, the team would either over-engineer the simple work or under-serve the complex work. Designed as a portfolio of four models matched to the cognitive shape of each step, the team produces a coherent end-to-end design that fits the work as it is.*

The aggregate occupational figures circulating in the wider conversation can obscure this kind of step-by-step variation. The World Economic Forum's *Future of Jobs Report 2025* (World Economic Forum 2025) projects a flattening of the task mix between humans, machines and hybrid configurations over the second half of the decade, and Chapter 30 returns to that picture in detail. Numbers at that altitude are useful for boards and headlines and almost useless for design. A single role, and often a single workflow inside that role, will contain tasks belonging to all three categories at once. The workflow map exists to expose precisely this task-level variation, so that different operating models can be applied within a single role rather than averaged across it.

Designing this portfolio without the map is impossible. With the map, the integration points fall out almost mechanically. The map tells the designer what kind of work happens where. The operating models tell the designer how AI participates in each kind of work. The two together produce a design rather than a collection of pilots.

14.6 The map is valuable before any AI is designed

A useful property of workflow mapping is that the map is valuable before any AI is introduced. Organisations frequently discover that the act of producing the map clarifies their own operations in ways that drive immediate improvement. Bottlenecks become addressable once visible. Implicit knowledge becomes shareable once articulated. Workarounds can be replaced by fixes once named.

This independent value matters for two reasons. The engagement produces something concrete before the AI design phase begins, which builds confidence and momentum. The organisation also has a document it can use for purposes beyond AI: training, succession planning, operational improvement, audit preparation.

Treating the map as a standalone deliverable disciplines the mapping process. If the map only has value in the context of AI design, the temptation is to capture only what AI design will need: the specifications, the decision points, the inputs and outputs. If the map must stand on its own, the mapper is forced to produce something complete, accurate and usable by people not thinking about AI at all.

This is a useful test. A good map is intelligible to the head of operations, the head of training and the head of audit, none of whom are reading it for AI purposes. If only the AI team can read it, it is not yet a map.

14.7 A good map is produced through triangulation, not interview

Effective workflow mapping does not rely on interviews alone. Interviews capture how people describe their work, which is reliably different from how they do it. Practitioners simplify, omit steps that have become automatic and describe the documented process even when their real process diverges from it. Interview-only maps are clean, plausible and frequently wrong.

Observation supplements interviews. Watching the work being done, ideally across multiple cases and multiple practitioners, reveals steps that were never mentioned because they were never noticed. Observation surfaces side channels, informal coordination and moments where someone pauses, frowns and applies judgement they would not have described as a decision.

Documentation review provides the third leg. The official procedures, the system configurations, the historical artefacts of the work, including completed files, audit trails and exception logs, ground the map in evidence. When interviews, observation and documentation disagree, the disagreement itself is informative. It usually points to the parts of the workflow that most need attention.

The conversation that produces a good map is iterative. Early drafts are returned to the practitioners who described the workflow. They correct, refine and frequently add elements that surface only when they see the workflow rendered. The final map is co-produced with the people whose work it describes, and they recognise it as accurate. The recognition test matters. A map the team does not recognise will not be used, no matter how detailed it is.

14.8 From the map to the design is a translation, not a leap

The transition from a completed workflow map to an AI integration design is the moment where the discipline of the previous phase pays off. With the map in hand, the designer can identify integration points, select operating models, specify human and AI responsibilities at each point and trace the information flows the new design will require.

This translation is rarely a single move. It proceeds through layers. The first layer identifies candidate integration points and proposes operating models. The second specifies what the AI does at each point and what data and context it requires. The third addresses the human side: how practitioners interact with the system, what authority they retain, what their work feels like under the new design.

Each layer is reviewed against the map. Does the proposed integration respect the actual handoffs, or does it require handoffs that do not exist? Does it draw on information that flows naturally to that point, or does it assume access to context that lives somewhere else in the organisation? Does the operating model match the cognitive shape of the work at that point, or does it impose a pattern the work resists?

The result of this disciplined translation is an AI integration design that fits the organisation rather than the other way around. The team continues to operate with structures it understands, augmented where augmentation creates value rather than replaced where replacement is easiest.

The map remains alive as the design evolves. It is the shared reference that keeps human and AI participation coherent over time, that surfaces drift between the design and the operating reality, and that grounds every subsequent design decision in something stronger than the memory of the people in the room.

The next chapter takes up the first concrete piece of that design: how AI automation is built once a workflow has been understood well enough to be built around. It is the discipline that turns the map's specification-heavy steps, the ones that bottomed out into clear rules and bounded inputs, into the connective tissue of an operating environment. Where the map said what the work is, automation is the practice that begins to determine what the work becomes once a system is built around it.

Chapter 15

Building AI Automations That Connect to Your Business

> *A distributed system is one in which the failure of a computer you didn't even know existed can render your own computer unusable.*
>
> — Leslie Lamport

15.1 Most of the value sits in the connections, not the model

A capable language model that cannot reach the claims database, the case management system or the customer record store is an island. Useful in conversation. Inert in the operating environment of the business.

The work of turning AI capability into operational throughput is not the work of the model. It is the work of the wiring.

This is the discipline that AI automation describes, and it is frequently confused with AI deployment. Deploying a model, subscribing to an assistant or rolling out a chat interface gives an organisation access to AI capability. Automation is something different. It connects that capability to the systems, data flows and processes that already run the business, and it lives almost entirely in the seams between them.

The distinction matters because it changes how the work is scoped. Teams that treat AI automation as a procurement decision underestimate the engineering effort by a wide margin. The model itself is often a small fraction of the build. Identity systems, source data, downstream workflows, audit logging and exception handling typically dominate both the timeline and the ongoing maintenance burden. The model arrives in days. The connections take quarters.

A working definition is useful here. AI automation is the practice of embedding AI capability into business processes such that work moves through those processes with reduced or eliminated human intervention at specific steps. The goal is operational, not experimental. The measure is whether the process now runs better, faster or more consistently than it did before, and whether it keeps running that way under load, under change and under scrutiny.

Two of the four operating models do most of the load-bearing work in this discipline. Knowing which is which prevents a great deal of expensive redesign.

15.2 Two operating models do most of the work here

The four operating models introduced in Part I describe different ways AI capability fits into work, distinguished by where decision-making responsibility sits. Automation primarily implements two of them.

Autonomous Agency describes AI operating independently within objectives and guardrails, taking actions and producing outputs without a human in the loop for each decision. Embedded Intelligence describes AI as a component inside automated infrastructure, invoked by other automated processes with no human interface during operation. These are the two models where automation lives.

A document classifier that runs whenever a new file lands in a shared inbox is Embedded Intelligence. A research agent that monitors a watchlist, gathers information and drafts briefings on a schedule is closer to Autonomous Agency. Both differ from Paired Cognition, where a human and AI work together in real time on the same problem, and from Task Delegation, where a human assigns discrete tasks and reviews each output.

The models are not mutually exclusive within a single workflow. A claims process might use Embedded Intelligence to extract structured data from an incoming document, Autonomous Agency to triage and route the claim within defined risk bands and Paired Cognition when an adjuster reviews the edge cases that the automation flagged. The design problem is deciding which steps belong to which model, and ensuring the handoffs between them are clean.

Choosing the wrong model for a step is one of the most common sources of failed automation. Putting Autonomous Agency where Paired Cognition is required leads to outputs that drift from intent, because the AI never built the shared mental model needed to keep its work aligned. Putting Paired Cognition where Embedded Intelligence would suffice creates a bottleneck that defeats the purpose of automating in the first place. A human now sits between every classification and the next system, and throughput collapses to whatever attention span the team has on a given afternoon.

Model selection happens before the build, and it cannot be retrofitted without redesigning the seams.

15.3 Not every process is a candidate, and the profile is recognisable

The decision about whether to automate at all is partly economic and partly architectural. Good candidates share a recognisable profile, and learning to spot that profile saves significant time and budget.

The single most important characteristic is regularity. Automations work best when the data they receive arrives in predictable shapes from predictable sources, and when the team also knows what a good output looks like for that input. A process that accepts a structured form, a known document type or a standard event payload, and that can be checked at the other end against a schema, a confidence threshold or a validation rule, is far easier to automate reliably than one that accepts free-form requests and produces outputs no downstream system can recognise as acceptable. Variability at the input is the single largest driver of brittleness in production. The absence of a usable definition of success on the output is what makes that brittleness invisible until it has compounded. The two travel together, and a candidate process needs both.

Regularity at the edges only earns its keep if the work in the middle can absorb the kind of variability the model itself produces. Language models give different outputs across runs even on identical inputs. Some tasks accommodate that gracefully, such as drafting a summary or proposing a category. Others cannot, such as calculating a regulated figure or producing a legally binding statement. The fit between the task and the variability of the model is a design constraint, not a tuning problem, and no amount of prompt engineering removes it. The same point governs what happens when the model is wrong, which is where blast radius enters the picture. A misrouted document that a human can re-route is a small problem. An incorrectly approved transaction that moves funds irreversibly is a much larger one. The combination of how much variability the work tolerates, and how contained the consequences are when the variability bites, governs both how much oversight the automation needs and whether the process should be automated at all rather than partnered.

A process that satisfies these conditions is automatable in principle. Principle, however, is not budget. The build only earns its keep at sufficient volume or repetition, because AI automation carries real engineering and ongoing operational cost. Processes that run thousands of times a day, or hundreds of times a week with material time savings each, are candidates. A process that runs twice a quarter rarely is, regardless of how attractive the technology looks in a demonstration. IBM's Institute for Business Value reported in 2025 that organisations treating intelligent automation as foundational rather than project-based were achieving cost reductions of up to 28 per cent at scale, with the gains concentrating among firms that had wired automation through their core processes rather than running it as a portfolio of isolated experiments (IBM Institute for Business Value 2025).

A useful summary sits underneath all of this. An automatable process is one where the work is regular enough at the edges to define, where the model's variability in the middle is acceptable, where the consequences of error are small enough to absorb and where the throughput is large enough to pay for the wiring. A process that meets every part of that description is a strong candidate. A process that meets most of it warrants careful design and probably a tighter scope. A process that meets little of it is usually better served by Paired Cognition or Task Delegation than by automation.

15.4 A handful of patterns recur across industries

Most AI automations in the wild are variations on a small number of patterns. Recognising them shortens the design phase considerably and helps teams reuse architectural decisions across multiple builds.

Document processing and routing is perhaps the most widespread. An incoming document, whether an insurance submission, a legal filing, a clinical referral or an invoice, is classified, parsed, validated and directed to the correct downstream queue or system. The AI components handle classification and extraction. The surrounding automation handles ingestion, validation, persistence and routing. The AI is the smaller part of the build.

Triggered analysis is a second pattern. An event occurs in a source system: a new contract is uploaded, a support ticket is opened, a transaction crosses a threshold. The automation invokes AI to analyse the event in context, produce a structured assessment and either store the assessment or pass it to another process. The human, if involved at all, receives the analysis rather than performing it.

Background classification covers the steady ongoing work of tagging, categorising and enriching records.

> *A mid-sized financial services firm runs background classification across its inbound communications. Every email arriving at the operations and service inboxes is tagged with intent, sentiment, urgency and the relevant client relationship. The classifications populate a workload view that the operations leads use the next morning to plan the day.*
>
> *None of the classification decisions are individually consequential, and none of them require partnership with a human. They simply need to be done consistently across tens of thousands of messages a week, which is unaffordable by hand and inconsistent if attempted ad hoc.*

Record maintenance and alert generation rounds out the common patterns. The automation reads from one or more sources, checks for conditions or anomalies and either updates downstream records or raises alerts to the appropriate human or system. These automations are the connective tissue that keeps multi-system environments coherent over time. They are unglamorous, persistent and often the highest-leverage work in the portfolio.

15.5 The integration is most of the work, and most of the cost

Teams new to AI automation often discover that the AI portion is not where the effort sits. The wiring is the work. Integration with existing systems usually dominates the build, and it dominates by a long way.

APIs must be authenticated against. Event triggers must be configured. Data formats must be mapped between systems that were never designed to talk to each other. Handoff points must be defined precisely, or work stalls mid-flight in places no one is monitoring.

This integration burden is not a technology problem so much as an organisational one. Each connected system has owners, change controls, security reviews and operational dependencies. An automation that touches five systems often requires alignment from five teams, each with its own priorities and its own definition of what "ready for integration" means. The pattern is familiar to anyone who has shipped enterprise software, and it does not disappear when AI is introduced. If anything, it intensifies, because AI automations frequently sit at the meeting points of systems that previously had no reason to talk.

The practical implication is that AI automation projects should be scoped with integration realism. Estimating only the model and prompt work produces budgets that are wrong by an order of magnitude. Including authentication, access control, error handling, retry logic, monitoring and downstream system changes produces estimates that hold up under delivery pressure.

A useful test is the connection inventory. Before designing the automation in detail, the team lists every system the automation will read from, every system it will write to, every event source that will trigger it and every channel through which a human might be notified. The length of that list is a leading indicator of the integration effort required. Most organisations are surprised by what the inventory reveals: typically a longer list than the original brief implied, and a tighter coupling between systems than anyone had documented.

The inventory is also where the integration risk concentrates. Each item on it is an external dependency. Each external dependency is a potential failure point.

15.6 Resilient automations are designed around the AI, not on top of it

What separates a resilient AI automation from a brittle one is rarely the model. It is the design around the model. Resilient automations assume that the AI will sometimes produce unexpected outputs, that source systems will sometimes return malformed data and that downstream systems will sometimes be unavailable. They handle these conditions explicitly, as part of the design, rather than treating them as exceptions to be discovered in production.

Validation gates are one common technique. Before acting on AI output, the automation checks that output against a schema, a set of business rules or a confidence threshold. Outputs that fail validation are routed to a human review queue rather than written to the system of record. This pattern preserves throughput on the easy cases while protecting against the long tail of failures.

A claims-handling automation in a general insurer extracts structured fields from incoming submission documents: policy numbers, loss dates, claimant identifiers, declared amounts. Each extracted field has a validation rule. Policy numbers are checked against the policy master. Loss dates must fall within cover. Claimant identifiers are cross-referenced.

Submissions where every field validates flow straight into the claims system. Submissions where any field fails validation are routed to a human handler with the failed field highlighted. The result is a system that handles the bulk of clean submissions automatically and surfaces the exceptions to people who can resolve them, rather than a system that either trusts the AI completely or distrusts it completely.

Idempotency and retry handling matter as well. Automations that may be triggered more than once for the same event need to recognise and handle duplicates without producing duplicate effects. Automations that depend on flaky downstream systems need backoff and retry strategies that do not amplify outages by hammering a recovering service. These are conventional engineering disciplines, and they apply to AI automation exactly as they apply to any other.

Edge case handling deserves explicit design rather than discovery in production. The automation should know what to do when it receives a document type it has not seen before, when extraction confidence is low, when a downstream system rejects its output and when a human override is requested. The list of edge cases is finite once the team thinks systematically about it. The cost of leaving them undefined is incidents, and incidents in AI automation tend to be expensive, because they often involve work that has already been actioned somewhere downstream by the time anyone notices.

15.7 Governance has to be built in when no human is watching

Automated AI processes raise governance questions that supervised AI use does not. When a human invokes the AI in real time, the situation in Paired Cognition or Task Delegation, oversight happens naturally through the act of using the output. The human is the first reviewer, every time.

When the AI operates autonomously inside infrastructure, no one is reviewing in real time. Oversight has to be specified in advance and wired into the system itself, because the human is not present to catch the failure as it happens. The governance question becomes a design question.

Monitoring is the foundation. The automation should emit telemetry about what it is doing, how often, with what success rate and at what cost. Dashboards and alerts make the behaviour of the automation visible to the team that owns it. Without telemetry, the automation becomes a black box that no one can reason about until it fails, and at that point, the conversation with the business is much harder than it needed to be.

Sampling and audit are companion disciplines. A subset of the automation's outputs should be reviewed regularly by a human qualified to assess them. The sample size depends on the blast radius and the variability of the work. The point of the review is not to catch every error but to detect drift, identify systemic issues and confirm that the automation is still operating within tolerance. Drift is the silent failure mode of AI automation, accumulating across runs without ever announcing itself.

Incident response completes the picture. When the automation produces a wrong outcome with material consequences, there should be a defined path for surfacing the problem, pausing the automation if necessary, investigating the cause and either remediating or rolling back. Treating AI incidents the same way mature engineering organisations treat production incidents is the appropriate posture. The unfamiliarity of AI does not justify a different operational standard. If anything, it justifies a slightly higher one, because the failure modes are less intuitive to reason about.

The accountability question sits beside the technical one. Someone has to own the automation. Not the team that built it and moved on, but a current owner with the standing to authorise changes, pause the system or take it offline. Automations without clear ownership tend to drift quietly until something visible goes wrong, at which point ownership is rapidly assigned, usually to whoever is least equipped to defend the prior design choices.

15.8 Several mistakes recur across organisations

A handful of failure patterns show up across organisations attempting AI automation for the first time.

Over-automating complex workflows is among the most common. Processes that require nuanced judgement, contextual interpretation of ambiguous inputs or genuine accountability for consequential decisions are usually better served by Paired Cognition than by Autonomous Agency. Forcing them into automation produces fragile systems that the business stops trusting, often after a single visible failure that becomes the story everyone tells about AI for the next 18 months.

Under-governing high-risk automations is the mirror image. An automation that touches regulated data, moves money or affects clinical or legal outcomes needs governance proportional to the stakes. Treating it as a productivity tool, with the same oversight one would apply to a desktop assistant, invites both incidents and regulatory exposure. The right level of governance is set by the consequences of being wrong, not by the convenience of being fast.

Ignoring failure modes is a third pattern. Teams sometimes ship the happy path and discover the unhappy paths in production. Mature automation programmes invert this sequence. They design for failure first and treat the happy path as the easy case. The cost of designing for failure up front is small compared with the cost of remediating in production while the business waits and the legal team asks questions.

The final recurring mistake is treating AI automation as conventional automation. The two share much of their engineering surface, but AI automation introduces variability that conventional automation does not have. Pretending that variability is a bug to be eliminated, rather than a property to be designed around, leads to expectations the system will never meet, and to a slow erosion of trust as the gap between expectation and behaviour widens.

Recognising AI automation as its own discipline, with its own design patterns and its own failure modes, is the foundation on which successful programmes are built. The model is the easy part. The connections, the governance and the design for failure are where the work really lives.

Building the automation is one thing. Building the team that can run alongside it, choose the right model for each piece of work and fold AI into the daily texture of the job, is another. That is the subject of the next chapter.

Chapter 16

Training Teams for Real AI Adoption

> *The distance between the actual development level as determined by independent problem solving and the level of potential development as determined through problem solving under adult guidance or in collaboration with more capable peers.*
>
> — Lev Vygotsky

When organisations roll out a new piece of enterprise software, training tends to follow a familiar shape. A vendor-provided curriculum. A two-hour session walking through the menus. A reference card pinned to a noticeboard. Users learn where the buttons are, which screens contain which features and how to complete the tasks the software was bought to support. This pattern works because most enterprise software is a faster way to do something the organisation already knows how to do. The work itself is unchanged.

AI breaks that pattern, and most training programmes have not yet noticed.

What changes when a team adopts AI is not which application they open in the morning. The interface, in most cases, is almost trivially simple: a single text box, perhaps a microphone icon, perhaps a sidebar that opens inside an existing tool. What changes is something far less visible. How cognitive work gets distributed. How thinking gets externalised. How decisions get made between a person and a system that can reason in language. Training that focuses on the interface teaches the smallest part of what matters. The largest part is left untouched, and adoption stalls precisely where it should be picking up speed.

Programmes also tend to be designed against the wrong baseline. McKinsey's 2025 *Superagency* survey (McKinsey & Company 2025a) reported that C-suite leaders estimated only 4 per cent of employees were using generative AI for at least 30 per cent of their daily work, while the actual figure from the same employees was closer to 13 per cent, roughly three times higher. The leadership view, in other words, was lagging the operating reality. Training built on the assumption that the team has barely begun will misjudge what the team already does, what it already needs and what it is already trying to teach itself in private.

This chapter argues that AI training is a different discipline from software training. The discipline is structured around two things: operating model literacy, and integration into real workflows. Tool familiarity is the residue of training done well, not its purpose. The most common failures in AI training programmes are predictable consequences of getting that orientation wrong.

16.1 Software training works because the working model holds still

Most professional skills are layered onto a stable working model. A solicitor learning a new case management system still practises law the same way. An underwriter learning a new pricing platform still assesses risk through the same conceptual lens. A finance analyst learning a new general-ledger module still closes the books in the same shape. The system changes while the working model holds steady. Training can therefore concentrate on the part that is new, the interface, and assume the rest is intact.

AI inverts that relationship.

The interface is often the least interesting part of an AI tool. A finance analyst who learns to use a generative AI assistant by asking it to summarise a document has not really learned to work with AI. They have learned a single transactional pattern. The deeper questions remain unanswered. When to use AI at all. What kind of conversation to have. How to express thinking that is not yet fully formed. How to integrate AI output back into a workflow that has its own structure and its own quality bar. The tool gets mastered without the discipline ever forming.

This is why traditional software training methods produce so little real adoption. Click-through demonstrations, scripted exercises and feature checklists assume the working model is stable and the only unknown is the interface. With AI, the working model itself is the thing being learned. The new vocabulary, the new habits of mind, the new judgements about which approach fits which kind of work, are not finishes layered onto an existing practice. They are the practice.

The shift is closer in character to learning a new professional discipline than to learning a new application. Organisations that recognise this from the outset design programmes that look more like capability development than tool rollout.

Their results look different too.

The training market has not caught up to the shift. The OECD's 2025 policy brief *Bridging the AI Skills Gap* (OECD 2025) found that roughly one in three job vacancies across OECD economies showed high AI exposure, while only between 0.3 and 5.5 per cent of training courses across Australia, Germany, Singapore and the United States delivered any AI content: a demand-supply gap of an order of magnitude. Organisations that wait for the external market to supply this discipline will wait a long time. The capability has to be built inside the work, by the team that is going to use it, in the context where it is going to be used.

16.2 A coherent capability model rests on operating model literacy

Effective AI training develops a single capability with several facets, but they are not equal. One sits beneath the others and decides whether any of them can do real work. The rest matter, but they matter in support of it.

That foundation is operating model literacy. A team needs to understand that there is more than one way to work with AI, and that different models suit different kinds of work. Paired Cognition suits ambiguous, judgement-heavy work where the human holds expertise the AI does not. Task Delegation suits well-specified, repeatable tasks. Autonomous Agency suits bounded execution within clear guardrails. Embedded Intelligence sits inside automated systems and rarely surfaces during operation at all, the territory of the next chapter. Trainees who cannot distinguish between these models tend to apply the wrong one and then conclude that AI is unreliable when in fact they have mismatched the approach to the task.

The misdiagnosis is the costly part. A team that has experienced an apparent AI failure rarely revisits the experience. They simply remember that AI did not work for that kind of work, and the assumption hardens into doctrine. Training that surfaces operating model literacy first prevents the misdiagnosis from forming. Almost every other training failure traces back to its absence.

A senior associate in a corporate law firm tries to use Paired Cognition for a stack of routine engagement letters and finds the process slower than drafting them himself. The next week he uses Task Delegation for a novel cross-border indemnity question and is surprised when the output reads as competent but unanchored.

The tool is the same tool in both cases. The mismatch is the problem in both cases. Training that names the four models and gives him practice choosing between them changes everything that happens next.

The next facet is externalisation, and it is harder than it sounds. Paired Cognition depends on the human partner expressing raw cognitive content, half-formed observations, partial intentions, working hypotheses, in speech or text without first organising it into polished form. Most professionals have spent careers learning to present finished thinking, and the instinct to edit before speaking interferes with the process. Speech-to-text helps, because spoken thought tends to be less structured than written thought, but the underlying shift is mental rather than mechanical. It is the willingness to be visibly mid-thought in front of a system that is going to do something useful with what is offered.

The third facet is situational awareness inside the work itself. Real cognitive work moves through phases, Exploration, Scoping, Planning, Execution, Validation, Refinement, Reflection, and the AI plays a different role in each. During Exploration it might surface options the human had not considered, whereas in Execution its job is to carry out a plan the two have built together, and by the time Reflection comes around it is helping the practitioner notice what worked, what did not and what should be carried forward into the knowledge base. Inside any given phase, the practitioner also needs to recognise which conversation they are in: Direct Conversation, where one human works with one AI; Observed Conversation, where humans dialogue with each other and the AI processes the transcript afterwards; or Assisted Conversation, where the AI participates in the room in real time. Phase and conversation type together describe the shape of the moment. Teams that read both and move between them deliberately get substantially more out of AI than teams stuck in Execution mode in a single chat window, which is the combination where AI output tends to be most generic.

The facets reinforce each other, but operating model literacy carries the others. A practitioner who has chosen the right model still needs to externalise to make Paired Cognition work, and still needs to read the phase and the conversation. A practitioner who has chosen the wrong model gets none of that back, however well they read the rest. Training programmes that lead with externalisation drills or conversation taxonomies and treat the operating models as background reading produce participants who can describe AI competently and apply it badly.

16.3 Paired Cognition deserves the most deliberate training design

Paired Cognition is the operating model that most rewards careful training, because it is the one most distant from how teams have historically used software. Although the interface looks familiar, the relationship sitting behind it is unlike anything the team has worked with before. Training that does not address this gap directly leaves participants imagining they have understood something they have only seen described.

The starting point is direct practice in externalising thinking under realistic conditions. Trainees need to work on actual problems from their own role, not staged exercises, not generic case studies, and they need to be coached through the experience of speaking unfinished thoughts aloud while the AI listens and reflects. The first time most professionals do this, it feels exposing. The second or third time, it begins to feel productive. By the tenth time, it begins to feel normal. None of this happens from a slide deck.

> *A consulting team adopting Paired Cognition spends its first training session not on prompts but on a single ninety-minute exercise. Each consultant brings an active engagement, dictates their current thinking about the problem for ten minutes without editing, and reviews what the AI produces in response.*
>
> *By the end of the afternoon, three of the seven have already moved a piece of real client work forward. None of them could describe a feature they had used. All of them could describe how the way they were thinking had shifted.*

The next focus is the construction of a shared mental model. The human and the AI need to converge on a common picture of what is being worked on, what matters and what counts as a good outcome. This rarely happens automatically. Trainees need practice noticing when the AI has drifted, when its picture of the problem has gone subtly wrong and how to repair the shared understanding without starting again from scratch. The skill is closer to working with a new colleague than to operating a tool.

The seven phases give structure to this practice. Trainees learn to recognise when they are in Exploration and should resist premature Execution; when they are Scoping and should narrow deliberately; when they are Planning and should make commitments explicit. The Reflection phase deserves particular attention, because most professionals are unaccustomed to using AI as a mirror on their own thinking. Done well, Reflection turns each piece of work into compounding learning that flows into the knowledge base. Skipped, it leaves each engagement isolated and the working memory of the team stranded inside individual heads.

Directed execution is the final piece. Once a shared mental model exists, the AI can carry out substantial work without a fresh brief, because the context is already in the room. Trainees need to feel the difference between handing the AI the cold brief and handing it a task within a relationship that has already been built. The quality gap is usually large enough to be self-evident once experienced, which is why training has to create the experience rather than describe it. Words on a slide do not do this work.

16.4 Failing AI training has a recognisable shape

Training programmes fail in patterned ways. Each pattern has a visible signature in the room and a predictable trail in the weeks that follow. Each is a consequence of treating AI like ordinary enterprise software, and one of them is responsible for more stalled adoption than all the others combined.

That dominant failure is interface-only training. The tell is the slide deck. Sessions cover prompts, settings and features, and the room fills with people copying example prompts into a notebook. Participants leave able to operate the tool and unable to integrate it into their thinking. Three weeks later the licences are still active and the usage logs are flat outside a handful of early adopters. The vendor's curriculum has been delivered in full. The working model has not been touched.

Beyond the absence of progress, interface-only training actively miscalibrates expectations. Participants come away thinking they have been shown what AI is, and what they have been shown is the chrome. When a colleague later describes what Paired Cognition feels like in practice, it sounds to them like an exaggeration of the demo they sat through. They have a mental model now, and it is wrong, and it took the most expensive day of the rollout to install it. Subsequent training has to undo the first session before it can do anything else. This is why the most common pattern is also the most costly: rather than a missed opportunity, it amounts to negative learning, paid for at premium rates, and it sets up the failure mode every other intervention has to work around.

Disconnection from real workflows is the next most damaging pattern, and it tends to compound the first. The signature is generic exercises producing generic skills. A finance team that learns AI through abstract examples will struggle to apply the same skills to month-end commentary, variance analysis or board reporting, because the connecting tissue between the training and their actual work was never built. The diagnostic is straightforward. Ask a participant a fortnight after the session what they are now doing differently with one of their own deliverables. A vague answer means the training did not connect at all, and a specific but trivial answer means it half-connected and stopped there.

An insurance underwriting team is taken through a half-day workshop using fictional scenarios designed by the training vendor. The exercises are well-constructed. The participants enjoy them. Three weeks later, when those underwriters sit down with a genuinely complex submission, they default to the same instincts they have always used.

The training did not connect to their real submissions and left no thread back to the work. A revised programme uses three live submissions from the team's own queue, redacted appropriately, and produces a different outcome inside the same number of hours.

A subtler failure is ignoring the knowledge base. The visible signature is a team that uses AI as a personal assistant rather than as one node in a system. AI works best when it has access to the organisation's own context in structured form, including the decisions, conventions and prior outputs that build up around any active practice. Training that treats the AI as a standalone assistant misses a large part of where the value comes from. Participants who have not been shown how their inputs feed the knowledge base, or how the knowledge base in turn improves future work, will never use it well. The third node of the partnership is invisible to them, and over the months that follow the team's AI work stays as good or as bad as the individual on a given day. What matters most is that the team's capability does not compound. A training programme that produces individually capable users but no compounding asset has built nothing the organisation can keep.

The one-off event failure barely needs examining, since AI capability does not embed in a single afternoon. Strong satisfaction scores on the day, almost no behavioural change six weeks later and the question of why that happens answers itself the moment it is asked. The remedy lives in the next section.

These patterns are the most reliable diagnostics for whether a training programme has been built around the discipline or around the tool. A programme exhibiting any one of them is producing trained users without producing capability. A programme exhibiting interface-only training is producing something worse.

16.5 Capability persists when training is part of how the work is done

Capability persists when training is integrated into the operating rhythm of the team rather than scheduled as a discrete event. New practices have to be embedded into briefings, reviews, handovers and retrospectives. People have to use AI on real deliverables under light supervision, with someone available to coach them through unfamiliar moments. Reference material has to be built that captures what good looks like in their specific context, not a generic best-practice guide, but a living record of how this team works with AI on this kind of problem.

The knowledge base plays a compounding role here. Early in adoption it is sparse, and AI work depends heavily on what each individual brings to each session. As the base grows, accumulating decisions, conventions, examples and prior outputs, the AI becomes more useful with less effort. Training benefits compound in parallel. Each new participant inherits the work of those who came before, rather than starting from scratch. The first cohort is doing more than learning. They are seeding the substrate that subsequent cohorts will inherit.

This is one reason the early period of adoption deserves heavier training investment than the later period. A finance function that trains its first ten analysts carefully, treats their early outputs as seeds for a growing knowledge base and embeds Reflection as a routine, will train its next fifty analysts with substantially less effort and better results. The investment is front-loaded by design.

16.6 Partnership framing reduces resistance, productivity framing hardens it

There is a useful pattern in how teams react to AI training, and it runs counter to the assumptions many leaders bring to the topic. Sceptical teams, when introduced to AI as a cognitive partner rather than as a replacement, tend to become enthusiastic, sometimes within a single session. The threat narrative dissolves once people experience AI as something that supports their thinking rather than something that competes with it. Adoption becomes something the team drives for itself.

Paired Cognition is the methodology with the least organisational resistance, for exactly this reason. It positions the human professional, with their judgement and expertise, as the centre of the work. The AI contributes, and the human leads. Teams in technology, professional services and finance have moved from active scepticism to active advocacy in the space of a few weeks once they have understood the partnership model in practice. What engineers the shift is not messaging but experience.

Training that frames AI as partnership therefore does double duty. It builds capability and dismantles resistance, in the same sessions, with the same exercises. Programmes that lead with productivity rhetoric or replacement rhetoric tend to harden resistance, even when the tools themselves are identical. The framing carried into the room shapes whether teams arrive at the next session with curiosity or with their arms folded.

16.7 Measuring real learning means looking past surface familiarity

Surface familiarity is easy to measure and tells almost nothing. Whether participants can describe a feature, recall a prompt pattern or rate the session highly is a weak signal of capability change. The leading indicators of genuine learning are different and more telling.

They include whether teams begin to externalise their thinking habitually rather than only when prompted. Whether they choose operating models deliberately, naming them as they go. Whether the knowledge base grows in quality and use. Whether AI shows up in real deliverables rather than only in training exercises. Whether Reflection appears as a routine rather than an afterthought. Whether the diversity of conversation types in active use expands beyond the Direct Conversation. Whether participants are willing to bring difficult, ambiguous problems to AI rather than only the simple ones.

These behaviours point to a working model that has actually shifted. Their absence, even alongside high satisfaction scores from the training itself, suggests the surface was touched and the substance was not.

The goal of training is adoption that lasts. That is a different objective from a successful training event, and it requires a different kind of programme. One designed around the working model rather than the interface. Around the capability beneath the tool rather than the feature list. Around real workflows rather than staged exercises. Organisations that hold this distinction clearly build AI capability that survives the transition from project to business as usual. Organisations that do not hold it produce trained users and stalled adoption, and spend the next budget cycle wondering why the gap between capability and outcome refuses to close.

Chapter 17

Embedding AI in Software Systems as Core Capability

> *There are two ways of constructing a software design: one way is to make it so simple that there are obviously no deficiencies, and the other way is to make it so complicated that there are no obvious deficiencies.*
>
> — C. A. R. Hoare

Training equips individuals to partner with AI on the work they own, but some capability cannot be carried by a person: it must be built into the systems those people depend on, present by design whether or not any trained individual is in the room.

There is a category of AI work the user never sees. It does not appear as a chat interface, a copilot panel or a button on a screen. It runs inside the plumbing of a software system, invoked by other software, producing outputs that feed into subsequent automated steps. The human role is confined to design and oversight, while at runtime the AI is just another component doing its job.

This is the territory of Embedded Intelligence, and it is where AI moves from being a tool people use to being a capability that systems possess.

When AI is treated this way, it stops being a feature added to an application and becomes a component of the architecture. It carries the same design weight as a database, a queue or an authentication service. The decisions about where it sits, how it is invoked and how it fails are decisions about the shape of the system itself. Those decisions tend to outlast the team that made them.

That permanence is worth taking seriously. The choices made in the first weeks of an embedded AI design will shape the cost of every change for the next several years.

17.1 Designed in, or bolted on?

The choice that shapes everything else happens early. An organisation can design a system with AI components from the outset, or it can attempt to retrofit AI into a system built without them in mind. The two starting points produce architectures that diverge sharply within months, and the second is almost always more expensive and more fragile than it appears at the time.

Systems designed with AI in mind treat language model calls, vector lookups and inference services as first-class architectural elements. They draw an explicit boundary between deterministic logic and probabilistic logic, and they refuse to mix the two casually. Their queues and timeouts are shaped to the latency profile of model calls. Their monitoring distinguishes a slow model from a wrong model from an unavailable one.

Retrofits, by contrast, tend to insert AI calls into code paths that were never designed to wait, fail unpredictably or produce outputs of variable quality. The result is a system that works in demonstration and breaks under load.

> *A claims platform built to process structured forms in milliseconds does not gracefully accommodate a synchronous call to a generative model that takes several seconds and occasionally returns malformed output. The first time the model provider has a bad afternoon, the platform stops processing claims. The team adds a timeout. The next time, the timeout fires before the legitimate response arrives, and the claim is silently dropped.*
>
> *Each fix patches the symptom. None of them address the underlying mismatch between an architecture designed for deterministic processing and a component that is anything but.*

The lesson is not that retrofitting is impossible. It is that retrofitting should be approached as a redesign rather than an insertion. The places where AI lives in a system are structural, and pretending otherwise produces architectures that are tense in ways their designers never intended.

The distinction sets the budget for everything that follows. Design it in: more cost in the first quarter, less in every subsequent year. Bolt it on: less now, more later, year after year. Both decisions are recoverable. Only one compounds in the right direction.

17.2 Embedded Intelligence is the operating model that governs AI inside the architecture

Embedded Intelligence is the operating model that governs most AI integrated directly into software. The decisive axis here, as elsewhere in the framework, is decision-making responsibility, and Embedded Intelligence collapses runtime responsibility entirely into design time. The defining property is that AI functions as a system component invoked by automated processes, with no human interface during operation. The human role is confined to designing the system, configuring its behaviour and reviewing outputs through whatever monitoring the team has put in place.

A typical example is a background service that performs authoring duties asynchronously. Tasks arrive on a queue, the service picks them up, calls a model, post-processes the result and writes it to a destination. Nothing about this loop involves a person at runtime. The same pattern extends to multiple independent services performing various behind-the-scenes duties, each one quietly doing its job inside the larger system.

What makes this model conceptually simpler than the others is that it is essentially traditional software architecture with AI components replacing conventional processing logic. Where a system might once have used a rules engine, a regex parser or a hand-coded heuristic, it now uses a model call. The interfaces around that call (inputs, outputs, error paths, observability) look much like the interfaces around any other piece of software.

This familiarity is both an advantage and a trap.

It is an advantage because experienced software architects already know how to design components, queues and pipelines, and most of that knowledge transfers directly. It is a trap because AI components have failure modes that conventional components do not. Treating a model call as just another function call leads to systems that work most of the time and surprise their owners the rest.

17.3 A new class of component, with unfamiliar failure modes

This is the point most often missed by teams arriving from a deterministic-software background, and the misunderstanding shows up later as production incidents.

A robust embedded architecture takes AI failure modes as design inputs rather than as edge cases to be discovered in production. A conventional component, given the same input, returns the same output, fails for understandable reasons and either works or does not. An AI component does none of those things reliably.

It times out unpredictably. It returns outputs that are syntactically correct but semantically wrong. It occasionally produces results that look plausible but contradict obvious facts. It drifts quietly when a provider updates a model behind a stable version label, so that a service which passed every test on Monday begins to misbehave on Wednesday for reasons no recent commit can explain. None of these are exotic failures. All of them happen in production systems at scale.

The trust gap this creates is now well documented. Stack Overflow's 2025 developer survey reported that 84 per cent of developers were using AI tools, but only 29 per cent trusted them, down from 40 per cent the year before. The most cited frustration, named by 45 per cent, was "AI solutions that are almost right, but not quite," with 66 per cent saying they were spending more time fixing such output (Stack Overflow 2025). Almost-right is the characteristic failure mode of a probabilistic component, and it is precisely the mode that conventional testing was never built to catch.

The discipline that absorbs most of the design budget is graceful degradation. When a model call fails, the system needs to know what to do. The fallback should be designed, named and tested before production, not improvised during the first incident. This is the work that distinguishes embedded systems that survive their first bad afternoon from those that do not, and it is the work most often deferred until an incident forces it.

> *For a fraud-scoring service, the answer might be to retry on a secondary provider with a tighter timeout, and if that fails, to fall back to a deterministic rules-based score that errs toward caution. The transaction still gets a decision. The decision is conservatively wrong some of the time, which is acceptable. The transaction never hangs, which is not.*

For a document extraction service, the answer might be to fall back to a smaller, faster model with a narrower extraction schema, then surface the partial extraction with a flag indicating that downstream consumers should treat it as provisional. The record still has structure. The structure is honest about its provenance.

For a content enrichment pipeline behind a marketing platform, the answer might be to skip the AI step entirely on failure and let the work proceed without enrichment. A reduced experience is better than a stalled pipeline. The marketing email goes out without the personalised hook rather than not going out at all.

Different workloads, different fallbacks. The unifying discipline is that each failure path is as deliberately engineered as the success path. The teams that take this seriously tend to be the teams whose AI features remain in production a year after launch. The ones who treat fallback as an afterthought tend to roll back features quietly within months.

Around graceful degradation sit two further disciplines that take less explanation but no less rigour. Quality assurance for AI components cannot rely on conventional unit tests. A function whose output varies in subtle ways across model versions, prompt revisions and input distributions cannot be pinned down by a handful of fixed cases. Teams running serious embedded systems build evaluation suites (representative inputs scored against expected outputs or quality dimensions) and run them automatically when models or prompts change. A drop in extraction accuracy from 94 to 91 per cent across releases should be visible before deployment, not discovered through a customer escalation three weeks later. Production monitoring extends the same logic into live operation: sample a fraction of live outputs, score them, feed anomalies back into the evaluation suite, repeat. None of this is conceptually exotic for any team that already runs observability on its services. It is, however, frequently absent.

The combination of designed-in fallback, pre-deployment evaluation and live monitoring is what lets an embedded AI system be trusted in environments where mistakes have consequences. A claims platform extracting loss details does not rely on the model alone. It relies on the model plus validation, monitoring and human spot-check tuned to the cost of an error. This apparatus is the price of admission for putting AI inside production software anywhere the work matters.

17.4 The shape of an embedded AI component is determined by how it is invoked

Three integration patterns dominate. Real-time inference handles requests inside a synchronous code path, where a user or another service is waiting for a response within strict latency bounds. Asynchronous processing places the AI call behind a queue, allowing the system to absorb variable response times without blocking the caller. Background agents are triggered by system events and run on their own schedule.

Each pattern places different demands on the architecture, and the right choice depends on the latency tolerance of the surrounding workflow and the cost of a delayed or missing result.

> *In financial services, real-time inference might score a transaction for fraud risk inside the payment authorisation flow. The model has perhaps 100 milliseconds to return a score before the authorisation either proceeds or fails.*
>
> *The architecture around the model is built for that budget: a co-located inference service, a feature store with sub-millisecond lookups, a fallback policy that defaults to a conservative score if the model is slow or unavailable. The integration is unforgiving because the surrounding flow is unforgiving.*

> *In legal operations, asynchronous processing might extract structured data from inbound contracts and write it back to a matter record. The user does not wait for the result, but the structured output must arrive before the next reviewer opens the file. The system tolerates seconds of latency where the fraud-scoring system would not tolerate one. It does not tolerate silently failing to enrich the record before the reviewer arrives.*

> *In insurance technology, background agents might triage incoming claims overnight, sorting them into queues by complexity and assigning preliminary tags. There is no user waiting and no immediate downstream consumer. The success criterion is that adjusters arriving in the morning find the work organised the way it would have been by an experienced operations lead working through the night. A delay of an hour is unremarkable. A systematic mis-tagging of high-severity claims is a serious incident.*

The three patterns differ in latency tolerance, in what counts as failure and in the architectural moves that follow. Real-time inference treats a missed budget as the failure; the architecture is built around the budget. Asynchronous processing treats silent non-arrival as the failure; the architecture is built around guaranteeing that something useful reaches the next step on time. Background agents treat systematic mis-classification as the failure; the architecture is built around evaluation, not throughput. In each case the choice of pattern sets what the rest of the design must answer to.

17.5 Which decisions reach far into the future?

Some decisions in this work are reversible and cheap. Others shape the cost of every change for years. Three categories sit firmly in the second group: model selection, vendor dependency and capability evolution planning.

Model selection is where many architectures pin themselves to a particular provider's interface, prompt format and feature set. The first version ships faster as a result. Every subsequent change pays for that decision. Tightly coupling a system to one provider produces faster initial delivery and slower long-term flexibility. Abstracting behind a common interface produces the opposite.

Most mature architectures end up somewhere in between. The most critical paths (the ones that handle the highest volume, sit closest to user-facing latency budgets or have the strongest regulatory exposure) are abstracted so that providers can be swapped without rewriting consumers. The experimental paths are tied directly to whatever provider-specific capability they need to function. The boundary between the two is itself a design decision worth making explicitly.

Capability evolution planning is the discipline of building systems that can absorb model improvements without being rebuilt. Models get better, faster and cheaper at a pace that few other components in a software system match. An architecture that pins itself to today's capabilities will be paying for that decision two years from now. An architecture that treats the model as a swappable component captures the upside as it arrives.

The teams that get this right tend to share a habit. They review their model choices on a regular cadence (quarterly is common) and treat the review as a first-class operational activity rather than a discretionary one. The review is short when nothing has changed and substantive when something has. Either way, it happens.

17.6 Other operating models appear at the edges of an embedded system

Embedded Intelligence is not the only model that lives inside software. Recognising which operating model is in play at which point in the system clarifies the design.

Autonomous Agency appears wherever a system needs more dynamic decision-making than a single inference call can provide. An agent that investigates an alert, gathers context from multiple sources, decides whether to escalate and writes a structured summary is operating with autonomy inside a software boundary. It needs objectives, guardrails and the same governance discipline as any other agent, not the lighter touch appropriate to a single embedded component.

Paired Cognition appears at the integration points where humans interact with the system. The platform itself runs on Embedded Intelligence, but the tools and interfaces it exposes to operators, analysts and decision-makers may be designed to support genuine cognitive partnership.

> *An underwriting platform might use embedded models to enrich every submission overnight, extracting risk attributes, matching against comparable accounts, surfacing unusual exposures. The enrichment happens in the architecture, invisibly. The enriched view is then presented through an interface that lets the underwriter think alongside the AI rather than merely review its output.*
>
> *The same submission flows through three distinct operating models on its way to a decision: Embedded Intelligence for ingestion and enrichment, Autonomous Agency for routine triage, Paired Cognition for the underwriter's reasoning on the ones that warrant it.*

The embedded components are infrastructure and should be governed as such. The agentic components need objectives and guardrails. The interactive components need to support externalisation, dialogue and shared mental models. Confusing these layers produces systems that are well-engineered for the wrong purpose.

17.7 Several integration mistakes appear repeatedly across organisations

Treating AI as an afterthought is the most common. A platform is built, ships and reaches stability. Only then does someone propose adding AI features. The result is a layer of model calls bolted onto an architecture that was never designed to host them, and the seams show.

Inadequate fallback design is a close second. Systems are deployed with the implicit assumption that the model will always respond and always respond well. When it does not, the system either stalls or produces nonsense, and operators discover the gap during an incident rather than during design review. Adding fallbacks after the fact is harder than adding them at the start.

The absence of production monitoring is a third. Teams without it cannot detect quality drift, prompt regressions or model deprecations until users complain. By the time the complaints arrive, the drift has been operating for weeks and the work to diagnose it spans several releases.

Ignoring latency implications is the fourth. User-facing flows that hang on synchronous model calls are a familiar pattern in retrofitted systems, where an asynchronous queue would have been a better fit from the beginning. The cost of unwinding that decision after launch is invariably higher than the cost of making it correctly during design.

Each of these mistakes has the same root: assuming that AI components behave like the components a team is already used to. They do not. They behave like a new class of component that happens to look familiar from the outside.

17.8 Embedded AI is disciplined software architecture applied to a new kind of component

The component happens to be probabilistic, occasionally unreliable and rapidly evolving. The engineering virtues it asks for are familiar ones: explicit interfaces, designed-in failure handling, observable behaviour, deliberate vendor decisions, an architecture that can change without being rebuilt.

Systems built with care in this style absorb new capability as it arrives. A better model becomes a configuration change rather than a project. A new provider becomes an option rather than a migration. A regulatory question becomes an audit log query rather than a forensic exercise.

Systems built without that care become harder to change with every model release. The gap between the two grows wider every year. The teams that took the architecture seriously in the first place compound their advantage. The teams that treated AI as a feature spend the next several years catching up to where they could have started.

The empirical picture supports this. DORA's 2025 *State of AI-Assisted Software Development* report, drawing on responses from nearly 5,000 technology professionals, found that AI adoption now correlates positively with both throughput and product performance, with the quality of the team's internal platform identified as the decisive variable (DORA 2025).

The disciplines of Part III have now covered strategy, mapping, automation, training and architecture. Architecture sets what a system can carry; the next chapter sets what an organisation can absorb. The last discipline is the pilot, where the design meets a chosen team on a Tuesday morning and the strategy starts producing evidence rather than argument.

Chapter 18

Piloting AI: How to Prove Value Before You Scale

> *The first principle is that you must not fool yourself, and you are the easiest person to fool.*
>
> — Richard Feynman

Until the pilot starts, every part of the design has been a hypothesis. Workshops have produced enthusiasm. Workflow maps have produced clarity. Training schedules have produced calendar entries. None of it is proof. The pilot is where the design discovers what the organisation can actually absorb, and the question that decides everything else is whether what worked on the whiteboard survives a real team's working week.

That last word does most of the work in this chapter. A pilot is not a proof of concept for the technology. The technology can usually be assumed to work, or shown to work in a smaller, cheaper exercise that does not require a redesigned workflow and a chosen team. A pilot answers a different and harder question.

It asks whether this organisation, with these people, on this workflow, can absorb a new way of working and keep doing it after the novelty fades.

18.1 The pilot exists to manage uncertainty about adoption, not capability

The instinct to treat a pilot as a capability test is widespread, and it is the source of more wasted pilots than any other single error. A team is given a tool, a problem and a few weeks. The tool produces outputs. The outputs look impressive. The pilot is declared a success and the rollout begins, at which point the organisation discovers that producing impressive outputs in a sandbox tells it almost nothing about producing useful outcomes in production.

The capability question matters, but it is rarely the bottleneck by the time a serious pilot is being planned. The strategy work done earlier has already filtered for workflows where the technology is plausibly capable. The workflow mapping has already located the cognitive shape and the integration points. By the time a pilot is on the table, the question is whether this team, embedded in this department, with this management line and this knowledge base and these data flows, can absorb the new operating model and still produce work the organisation trusts.

Because that is an organisational question, it cannot be answered by demonstration, only by trying.

A well-run pilot does several things at once. It contains the blast radius if the design is wrong, because failure on a bounded workflow is recoverable in a way that failure across a department is not. It surfaces the small, practical, infuriating issues at a scale where they cost a fortnight of adjustment rather than a quarter of remediation: the photocopier-jam problems that no architecture diagram predicts and that no executive ever sees until they are causing missed deadlines. And it generates evidence specific enough to support a real scaling decision rather than a rhetorical one.

The confidence-building deserves particular emphasis. People become advocates for what they have personally seen working. A pilot that visibly succeeds inside a respected team produces internal champions whose endorsement carries further than any vendor case study or executive memo. That kind of credibility cannot be bought.

18.2 Choosing the pilot workflow is the decision that decides the pilot

Pilot selection is where most pilots are won or lost, long before any work begins. When the workflow has been chosen badly, no amount of operational care will rescue the pilot, whereas a workflow chosen well forgives a great deal of clumsiness later.

A candidate workflow has to earn its pilot. Several conditions need to hold at once, and a workflow failing on any of them should be replaced rather than nursed through to a conclusion the evidence cannot support.

The first thing to ask is whether success would actually matter. A finance team piloting AI-assisted variance commentary is choosing well because variance commentary consumes meaningful senior time and produces meaningful insight. Piloting AI on a quarterly task nobody minded doing in the first place is choosing badly, regardless of how cleanly the pilot runs. Closely related is scope: the workflow needs to be small enough to redesign and observe in detail, with a defined boundary that distinguishes what is in the pilot from what is not.

> *A general insurance team piloting AI-assisted claims handling might place straightforward property claims under a defined value threshold inside the pilot, and leave complex bodily injury and large commercial losses out of scope. The boundary is not arbitrary. It marks the territory across which the pilot's findings will actually be valid.*
>
> *Findings about high-volume, low-complexity claims do not generalise to bodily injury, and pretending they do is how successful pilots produce failed rollouts.*

Success criteria need to be agreed before the pilot starts, by the people who will judge it afterwards. Vague criteria almost always resolve into political readings of ambiguous evidence, which is how the same pilot ends up celebrated by its sponsors and dismissed by its sceptics. The blast radius needs to be low: a law firm should not pilot AI on its highest-stakes litigation, but on contract review or research synthesis, where existing review processes will catch errors and the cost of a misstep is bounded.

The condition most often overlooked is operating model fit. The workflow being piloted should match the operating model the pilot is intended to test. Piloting Paired Cognition on a workflow that is genuinely Task Delegation in nature will produce confused results that flatter neither the model nor the team. The strategy and mapping work from earlier chapters has done the load-bearing analysis here. The pilot should respect it.

Taken together, these conditions describe a single workflow: one that matters, that can be cleanly observed, that can be judged honestly, that fails safely if it fails and that fits the operating model on trial. The discipline is to refuse a pilot until a workflow can satisfy them, and to keep refusing until one can. Most organisations refuse far too late.

18.3 A pilot designed in vague terms produces vague evidence

Once the workflow is chosen, the pilot needs to be designed with specificity. Vague pilots produce vague evidence, and vague evidence cannot support a confident scaling decision.

Several decisions sit upstream of the work and have to be made before it starts. Success criteria need to be both quantitative and qualitative, because a purely quantitative pilot will miss the texture of adoption and a purely qualitative pilot will miss whether anything actually changed. The operating model under test should be named explicitly: not "AI for credit memos" but Paired Cognition for credit memos, or Task Delegation for engagement letter drafting, or Embedded Intelligence for transaction monitoring. Naming the model focuses the design and prevents the team sliding between models in ways that confuse the resulting evidence.

> *A bank piloting Paired Cognition for credit memo preparation might define quantitative success as a measurable reduction in time from initial submission to credit committee, and qualitative success as evidence that credit officers find the AI's reasoning trustworthy enough to use without prompting from a sponsor.*
>
> *Either measure on its own would mislead. Time saved by officers who quietly stopped using the system after week three is not time saved. Trust reported by officers who never produced a faster memo is not adoption.*

Team selection matters more than organisations usually allow. The pilot team should be representative enough that the findings will generalise, and willing enough that the work will actually happen. A team instructed to participate without being persuaded will produce evidence about resistance, not about the workflow. A team of unusual enthusiasts will produce evidence that overstates likely results at scale, and the rollout will spend its first six months wondering why the magic has gone. A well-designed pilot, from inside, feels less like a launch and more like the team has been quietly given permission to think about its own work, slightly more reflective than usual, with someone in the room actually writing down what changed and why.

Duration and measurement bracket the rest. Most pilots benefit from a window of 6 to 12 weeks: shorter than that and the team is still learning when the pilot ends; longer and the pilot becomes its own permanent state, losing the urgency that drives Reflection. The measurement approach has to be settled before the work begins, with the team knowing in advance what data will be collected, how and by whom. Retrofitting measurement onto a pilot already running rarely succeeds. Baselines, once missed, are gone.

18.4 Pilots fail in predictable ways and the patterns are worth recognising

Gartner forecast in 2025 that more than 40 per cent of agentic AI projects would be cancelled by end-2027 due to escalating costs, unclear business value or inadequate risk controls (Gartner 2025b): the same pattern as the earlier generative-AI proof-of-concept abandonment trend, intensified. The failure modes are concrete and they recur across industries. Naming them does not make them avoidable, but it does make them recognisable when they begin to appear.

Scope set too broad is the most common error. A pilot that tries to test three operating models across five workflows in two departments over two months is not a pilot. It is a small transformation programme without governance, and it will produce evidence too entangled to interpret. The team will be busy. The deliverables will ship. The retrospective will be a meeting in which six senior people each describe a different pilot, agree it went well, and walk out unable to name what specifically should be repeated. Nobody will be able to say which workflow benefited from which model, or whether the gains came from the AI or from the simple fact that five workflows received executive attention for the first time in a decade.

Wrong success metrics is the second pattern. A pilot measured purely on cost reduction will miss the quality, satisfaction and learning effects that often matter more in early adoption. A pilot measured purely on adoption rate will miss whether the adopted workflow is actually producing better outcomes. The metrics must match the workflow's real value drivers, not whatever was easiest to count.

Absent Reflection is the third and quietest failure. The pilot ends, the deliverables ship, and the team moves on. Nobody pauses to capture what was learned. The organisation has technically done the pilot but cannot tell anyone what the pilot showed. The cost was incurred without the benefit, and the rollout will repeat from approximately zero.

The fourth pattern is the failure to capture both what worked and what did not. Honest pilots include the friction, the workarounds and the cases the AI handled badly. Without that texture, the scaling decision rests on incomplete evidence, and the rollout discovers the missing texture the hard way.

> *A technology firm piloting Paired Cognition for architectural decision-making produced a triumphant final report. Adoption was high, time to design proposal had fallen, and the senior architects were enthusiastic.*
>
> *The report did not mention that two of the four pilot teams had quietly developed a workaround for the knowledge base because nobody had time to maintain it, or that the Reflection phase had been skipped after the first three weeks because the pilot lead found it awkward to run.*
>
> *The rollout inherited the workaround as a problem and the absent Reflection as a habit. By month four, the rollout teams were using AI as expensive autocomplete, and nobody could explain when it had stopped being a partnership.*

Each of these patterns shares a root. The pilot was treated as a piece of work to complete rather than as an instrument for organisational learning, and its evidence-producing function was traded against its delivery-producing function whenever the two competed.

18.5 The Reflection phase inside a pilot is the most important output

The most important output of a pilot is not the deliverables it produces or the metrics it records. It is the learning the organisation extracts from having run it, and that learning is produced by the Reflection phase or it is not produced at all.

Of the seven phases, Reflection is the one that converts a pilot from an episode into an asset. The discipline of Reflection at workflow scale was developed earlier in the book. A pilot is the same discipline operating one level up. The deliverables are useful and the metrics are useful. The lessons are what make the pilot worth running.

A pilot's Reflection covers familiar territory, scaled up to match the larger frame. What decisions were made during the pilot that future decision-makers need to understand. What documentation needs to be updated, revised or created. What should be added to the knowledge base so that the rollout teams inherit it rather than reinvent it. What follow-up tasks should be assigned, and to whom. Where did the team encounter weak knowledge or unclear process that the pilot exposed but did not resolve. What workflow improvements emerged that should be carried forward. Who in the organisation needs to be told what the pilot found, and in what form.

Reflection is distinct from Completion, and the distinction matters more here than in any individual workflow. Completion closes the pilot administratively. The team is debriefed, the deliverables are filed, the budget is reconciled, the steering committee receives its final report. Completion is necessary and largely mechanical. Reflection is the harder, slower act of asking what the pilot taught the organisation, and ensuring the answer survives the team disbanding and the sponsors moving on to the next priority.

The pattern of doing Completion well and skipping Reflection is so common, and so quietly destructive, that protecting Reflection should be designed into the pilot from the start. Time should be budgeted for it. Attendees should be specified. Outputs should be defined. Without that structure, Reflection is the first thing sacrificed when the pilot runs long, and the pilot almost always runs long.

A pilot that completes without reflecting has consumed the budget without producing the asset.

18.6 A pilot proves something specific and nothing more

It is worth being precise about what a pilot proves. A pilot proves value in a specific workflow, in a specific organisation, with a specific team, over a specific period. That is all it proves. Deloitte's January 2026 *State of AI in the Enterprise* survey of more than 3,000 leaders, conducted in late 2025, reported that only 25 per cent of organisations had moved 40 per cent or more of their generative AI pilots into production, and that 54 per cent expected to reach that threshold within the next three to six months (Deloitte 2026a), meaning even the leaders' implicit forecast was that the bulk of their experiments would not scale.

It does not prove that the underlying technology works in general. That is a different exercise, often called a proof of concept, usually run earlier and at smaller scale. It does not prove that the operating model will succeed in other workflows. It does not prove that other teams in the organisation will adopt the same way as the pilot team did.

This precision matters because pilots are routinely asked to carry more evidence than they can support. A successful Paired Cognition pilot in a finance team's variance analysis tells the organisation a great deal about variance analysis, and a fair amount about that finance team. It tells the organisation rather less about whether the same model will succeed in a different function, with different work, different data and different culture. Treating the pilot's evidence as broader than it is leads to confident rollouts into workflows the pilot never tested, and to puzzled retrospectives six months later when the magic has not generalised.

The honest reading of a pilot is narrow. The narrow reading is also the useful one.

18.7 The transition from pilot to rollout is its own design problem

Several things change at scale, and the rollout has to be designed for them rather than assumed away. The pilot team was selected for representativeness or willingness, while rollout teams will not be. The pilot enjoyed close attention from leadership and external partners, while the rollout will compete for attention with everything else the organisation is trying to do. The pilot ran in a defined window, while the rollout has to become permanent practice.

The rollout design should preserve what made the pilot work, which means understanding what made the pilot work in the first place. This is where the Reflection output earns its keep. The decisions that produced success can be carried forward only if they were captured. The patterns that produced friction can be addressed only if they were named. Without that record, the rollout starts from approximately where the pilot started, and the pilot's investment is discarded.

A useful pilot also functions as internal communication. The team that ran it becomes a credible source of testimony for the teams that will adopt next, and the artefacts it produced (the redesigned workflows, the populated knowledge base entries, the changes in working practice) are visible evidence that something real happened. The organisation needs to see the new work happening before it will believe in it.

18.8 A pilot that underperforms still produces evidence

Not every pilot succeeds, and the response when one does not should be diagnosis before abandonment. A pilot that misses its success criteria has produced evidence. The evidence is worth understanding before the work is set aside, and the diagnosis questions are familiar from any piece of work that did not produce its expected result.

Was the workflow misjudged, or the operating model. Was the team the wrong fit, or was the team's resistance signalling something the design missed. Were the success criteria reasonable, or were they hopeful. Did the data and tooling support what the workflow required, or did the foundations break in ways that were not anticipated. Was the duration long enough to move past novelty, or short enough to retain urgency.

A pilot that underperforms because the operating model was wrong tells the organisation something important, and should not be read as a failure of AI. A poorly chosen workflow tells it something quite different, and a team set up to fail tells it something else again. What pilot failure usually looks like in the room is not dramatic refusal. It is a quiet drift in which the AI is opened less often each week, the sponsor stops asking about it in standups, and somebody eventually mentions it in a retrospective with the past tense already in their voice. The point of diagnosis is to extract the specific lesson from that drift, so the next attempt is informed rather than defensive.

The goal, finally, is adoption that lasts. Pilots are how organisations earn the confidence to commit to that adoption, and the evidence that the commitment is well placed. Run carelessly, they consume time and produce noise. Run with discipline, they are the most economical form of organisational learning available, and the foundation on which durable AI adoption is built.

Part III closes here. Strategy decided where to begin. Workflow mapping decided what to redesign. Automation, training and embedding decided how the new work would be supported. The pilot decides whether the redesign actually fits the organisation that has to live with it. These six disciplines are how an operating model framework stops being an idea and becomes practice.

The next part of the book turns from method to application, where the four operating models stop being abstractions and become the specific shapes that compliance, underwriting, legal advice, clinical reasoning, marketing and communications, engineering and teaching take when AI is brought into them. Part IV walks through seven industries in turn, beginning with the one whose accountability structures make the operating model distinction matter most: finance.

Part IV

Across Industries

The framework applied across seven sectors: finance, insurance, legal, clinical medicine, marketing and communications, technology and education.

A note on how to read this part. Seven chapters are a substantial commitment, and the part is not designed to be consumed exhaustively by every reader. The framework's character is visible in any one or two of the chapters in depth; reading the chapter that maps your own industry, then dipping into one or two others where the structural shape interests you, is enough to feel how the four operating models redistribute across different kinds of work. Reading the part end to end is welcome but not required, and skimming the industries you do not work in is not a failure of attention. The argument the rest of the book rests on is intact whether you read three chapters here or all seven.

Chapter 19

AI as a Cognitive Partner in Finance

Credit means that a certain confidence is given, and a certain trust reposed. Is that trust justified? and is that confidence wise? These are the cardinal questions.

— Walter Bagehot

A credit decision is rarely a calculation. It is an argument, made under pressure, that the obligor's circumstances, the portfolio's existing exposure, the market's current temper and the firm's stated risk appetite can be reconciled in favour of a particular outcome. The numbers form part of the argument, but they do not finish it. Whoever signs the memo is putting professional judgement on the line, and they will be asked to defend it later by an internal credit committee, by an auditor, by a regulator, by a client whose facility has just been declined.

Finance is past pilot. The Bank of England and FCA's joint survey of UK financial services firms (Bank of England and Financial Conduct Authority 2024) found that 75 per cent were already using AI, up from 58 per cent two years earlier, with adoption highest in insurance (95 per cent) and international banks (94 per cent), and roughly a third of all use cases delivered through third-party providers. The question facing the industry is no longer whether to adopt but how to govern what has already been adopted.

What separates the work that one financial professional does from another is rarely the subject matter. It is the density of judgement the work demands per hour. A wealth manager defending an asset-allocation shift to a fiduciary client is operating at the top of a spectrum. A reconciliations clerk matching settlement breaks against an exception report is operating near the bottom. Both are finance. Both are valuable. Neither is the same shape of work, and neither rewards the same operating model for AI.

This chapter walks down that spectrum. It begins where judgement density is highest, where the institution's accountability is most personally borne and where the consequences of misapplied AI are most expensive, and it ends where judgement has been so thoroughly codified into procedure that the work is properly invisible. Finance already thinks in terms of delegation, oversight, accountability and audit. The four operating models give that thinking a sharper vocabulary.

What follows is an argument, not a tour. The case for Paired Cognition in finance rests on a quality argument rather than a productivity one, and the spectrum is the clearest way to see why.

19.1 The top of the spectrum: where the institution's name is on the memo

What regulators reward is not speed but defensibility. A credit decision that can be explained, audited and justified against the firm's stated risk appetite is the goal, regardless of how long it took to arrive at it. The firm whose memos consistently anticipate the supervisor's questions before they are asked operates in a different regulatory posture from the firm whose memos answer those questions only on request. The strongest argument for Paired Cognition in finance has almost nothing to do with how fast the work gets done.

Faster compliance work is useful. A compliance function whose conclusions are demonstrably better grounded, more thoroughly cross-referenced and more consistently reasoned is something else entirely. The same shift applies across credit analysis, investment research, regulatory engagement and risk assessment. The point of partnership in this industry is not the hour saved. It is the depth of the analysis the human is able to bring to the question, and the defensibility of the conclusion that emerges from it.

This reframing addresses the most common professional concern about AI in regulated industries. Senior practitioners worry, reasonably, that AI will dilute the quality of judgement, introduce opaque reasoning into auditable processes or substitute statistical pattern matching for the careful, defensible analysis their fiduciary and regulatory obligations demand. The concern is well placed. Each of those failure modes is what happens when AI is deployed under the wrong operating model. Properly understood, Paired Cognition does the opposite: the human remains the decision-maker while the AI extends the depth and breadth of the analysis the human is already doing. The resulting work product is more defensible, not less.

There is a second-order benefit that finance teams routinely underestimate. The interaction between practitioner and AI in a Paired Cognition workflow produces an artefact that conventional work does not. A working session leaves a record. What was considered. What was rejected. Which alternatives were tested. Which precedents were checked. Which interpretations were ranked against which others, and on what basis. The conclusion that emerges sits on top of a visible reasoning trail, because the trail is a natural by-product of how the work was done.

For firms operating under model risk management frameworks and supervisory expectations around explainability, this is an asset rather than a burden. The audit trail in a Paired Cognition workflow is structurally richer than in a conventional one, provided the workflow has been designed with that documentation in mind from the outset. Reconstructed under pressure once the regulator asks, it is worth far less.

This is a distinctive claim and worth holding onto. Compared with a delegate-and-review workflow, where the analyst's reasoning is largely internal and the review produces a tick rather than a record, partnership produces evidence as a side effect of the work itself. The work that is hardest to defend in front of a supervisor is precisely the work where the reasoning happened in someone's head and was never written down. Partnership writes it down because that is how partnership operates.

Finance has been moving toward more demanding standards of explainability for a decade. The operating model that produces the richest evidentiary trail for the most consequential decisions is not a footnote in the case for Paired Cognition. It may be the central argument.

19.2 What partnership looks like in the work itself

Risk assessment. Compliance interpretation. Investment research. Credit analysis on non-standard exposures. Regulatory engagement. The common shape is sustained engagement with ambiguous material, where analyst and AI together build a shared understanding of the problem and neither can fully proceed without the other.

This is partnership in the strict sense the framework calls for. Not a brief handed off and a draft returned. A working session in which the human externalises context, the AI organises and surfaces, the human exercises judgement on what matters and the loop repeats until the conclusion has been reasoned through rather than merely produced.

Consider what this looks like in compliance work.

> *A senior compliance analyst at a mid-tier bank receives a new piece of regulatory guidance running to 40 pages. It references three prior rulings, cross-cuts seven internal policies and uses terminology whose meaning has shifted across successive interpretations. Working alone, she spends the better part of two days locating the relevant precedents, mapping the guidance against existing policy and assembling a picture of what is actually required. The mechanical labour of cross-referencing dominates. The contextual judgement that justifies her seniority is squeezed into whatever hours remain at the end.*
>
> *Working in partnership, the same analyst surfaces cross-references and implications inside an hour. The AI maps the guidance against existing policy, flags inconsistencies between the new requirement and three current procedures and proposes a structured set of interpretations ranked by how closely each tracks recent regulator commentary. She then exercises the judgement only she can: reading the regulator's likely intent against recent enforcement posture, weighing which interpretation best serves both compliance integrity and commercial reality and taking professional responsibility for the recommendation that goes upstairs.*

Neither party disengages. The output is qualitatively better than either could produce alone, and the analyst's name on the recommendation carries the same weight it always did. Arguably more, because the reasoning trail behind it is richer.

The same pattern holds in investment research. An analyst building a thesis on an industrial issuer must integrate financial statements, sector commentary, competitive positioning and macroeconomic context, then arrive at a view that holds together under interrogation by a portfolio manager who will ask the questions the analyst has not yet thought of. AI in partnership structures the universe of inputs, surfaces anomalies in the financials, proposes comparables and tests them, drafts the framing of the argument. The analyst contributes the contextual judgement: why this management team's recent missteps are recoverable, why the competitive moat is more defensible than the consensus assumes, why the macro overlay matters more here than in adjacent sectors. The thesis that survives this loop is not the thesis the analyst would have produced alone.

Directed execution, the mechanism by which a shared mental model produces output shaped to the actual matter, is what does this work. The shared mental model the analyst and AI build together is what allows the AI to draft sections of the thesis the way the analyst would have drafted them, only with broader reference and faster iteration. A cold brief produces a competent draft. A directed execution produces a draft that already reflects the position the analyst is taking.

19.3 Down a step: the everyday production work that surrounds the judgement

Partnership is not the right model for everything a senior practitioner does. A great deal of professional finance work is well-specified production: the substantive call has already been made, and what remains is to render it into a deliverable.

A portfolio manager asks AI to draft the client letter summarising the quarter once the view has already been formed. A credit analyst asks AI to extract covenants from a loan agreement she has read but does not want to re-read. A compliance officer asks AI to summarise a long document so she can decide whether to read it in full. There is no shared mental model to build, no sustained context to maintain. There is a clear instruction, a deliverable and a review, and for work the human already understands completely the overhead of partnership is unjustified.

The boundary worth defending is between drafting tasks where the practitioner can recognise quality on review and judgement tasks where review alone is not a sufficient quality control. A credit memo on a non-standard exposure is not a drafting task in this sense, even though it produces a written output. The reasoning that produces it requires the sustained engagement of partnership rather than the delegate-and-review rhythm of task work. The risk in mature finance teams is that Task Delegation becomes the default everywhere, including in places where it cannot carry the load.

19.4 Lower still: the operational work where agents earn their place

Below the production tier sits the high-volume operational work that fills the back and middle offices of every bank. Transaction monitoring. Sanctions screening. Anti-money-laundering alert triage. Reconciliation breaks. Customer onboarding checks. Document classification at the edges of the institution. Routine, well-specified, voluminous: the kind of work where consistency at scale matters more than any individual judgement.

Decision-making responsibility within scope sits with the agent. Accountability for the scope itself, its guardrails, its exception thresholds and the periodic review of agent performance, sits with the humans who designed and operate the system.

> *An AML triage agent works through the overnight queue of alerts generated by the firm's transaction monitoring system. It clears the bulk of clearly-benign alerts within defined parameters, escalates a smaller number with structured reasoning to human investigators and flags a third tier for second-line review where its own confidence is low or where the pattern matches one of the institution's recent typology updates.*
>
> *By morning, the human team is working a queue that has been pre-sorted by the same kind of judgement they would have applied themselves, only at a volume no team could reach. The investigators retain authority over every decision that matters. The agent has bought them the cognitive room to focus on the decisions that do.*

Finance has long operated this way with rules-based automation. AI-driven agents change what is possible inside the scope, with more nuanced classification, better contextual matching and more graceful handling of edge cases, but the basic shape is unchanged. The governance question is therefore familiar. How is scope defined. How is performance monitored. How are exceptions handled, and at what frequency does the institution sample the agent's decisions to confirm that the boundary it set still holds. Firms with mature operational risk frameworks adapt to AI agents more readily than firms without, because the conceptual machinery is already in place.

19.5 At the bottom of the spectrum: the work no one sees

Beneath all of it sits a layer where AI is invisible. No human user sits at runtime, and no agent operates against an objective. There is just a component in a pipeline, called by another system, returning a result and moving on.

Background analytical services score transactions, classify communications, summarise documents on ingestion, extract structured data from unstructured filings, enrich credit bureau information, tag risk factors across a research feed. The AI is treated as a vendor component or an internal service, with the usual concerns around versioning, reliability, monitoring and change management. Finance has substantial appetite for this model because so much of the industry's work depends on processing large volumes of structured and semi-structured information. The outputs feed into systems that humans or agents then act on, and the AI itself rarely becomes the focus of governance attention beyond standard model risk and vendor management processes.

The discipline here is design-time discipline. By the time the system is in production, the governance has already been done, or it has not, and the consequences will surface unevenly across the workflows that depend on it.

19.6 What the spectrum imposes: the constraints finance brings to the mix

A spectrum is useful only if the firm's governance structure can vary along it. Finance imposes constraints that other industries do not, and the operating model mix has to accommodate them rather than work around them.

Regulatory requirements vary by jurisdiction and product, but the common thread is auditability and explainability. Decisions must be reconstructable. Reasoning must be defensible. The chain of accountability must be clear, both for internal governance and for supervisors who may ask years later what was considered at the time.

Fiduciary duty raises the stakes further. A wealth manager acting on behalf of clients cannot delegate the duty itself, regardless of how the work gets done. AI can support the analysis, draft the recommendations, structure the reasoning, but the duty rests with the human professional, and the operating model must reflect that. Partnership aligns naturally with fiduciary work because the human remains the decision-maker throughout; agents and embedded systems are kept clear of decisions that carry fiduciary weight.

Model risk management adds a further layer. Any AI system whose outputs influence material decisions falls within scope, and firms must demonstrate that they understand the model's behaviour, monitor its performance and have controls for failure modes. The control surface differs by operating model: per-instance human review where there is one, framework-level governance where there is not.

The same Bank of England and FCA survey (Bank of England and Financial Conduct Authority 2024) reported that 84 per cent of firms had allocated AI accountability to a named individual, but only 34 per cent claimed complete understanding of the AI they were deploying, and 46 per cent reported only partial understanding. Accountability has moved faster than capability. A named owner whose grasp of the underlying system is partial is precisely the failure mode the operating model framing is designed to address: governance that names a person but cannot specify the kind of AI use being governed produces an audit trail without a defence.

These constraints are not obstacles to AI adoption in finance. They are the reason the operating model framing matters here more than almost anywhere else. Without it, governance teams default to one of two failure modes: refusing all AI use because the regulatory exposure feels unbounded, or approving it indiscriminately because no one has articulated what kind of AI use is being approved. With it, governance can vary by model, with controls calibrated to the actual risk surface of each.

19.7 The mix that a well-run financial services firm arrives at

A mature financial services organisation runs all four models simultaneously and deliberately, because the work itself is distributed across the spectrum. Front-office investment professionals, credit teams, compliance functions and risk teams partner with AI on their most consequential work, and delegate the everyday production tasks that surround it. Operations, customer servicing and surveillance functions deploy agents within carefully governed scopes, with clear escalation paths into the human teams that own the underlying risk. Technology and data functions embed AI throughout the analytical and reporting infrastructure.

The mix evolves as capability matures. The structural logic, matching the operating model to the density of judgement the work demands, does not.

Firms that get this right do not treat AI as a single capability to be procured and deployed. They treat it as a set of operating models to be matched thoughtfully to the work, with the governance, accountability and quality expectations appropriate to each. The professionals who learn to work in genuine partnership with AI on the hard problems, while letting agents and embedded systems handle the rest, are the ones whose institutions will set the standard for what regulated, fiduciary work looks like in the decade ahead.

What that standard offers is defensibility rather than speed, and in finance the difference between those two is anything but subtle.

Chapter 20

AI-Augmented Underwriting and Claims in Insurance

> *The volume and complexity of what we know has exceeded our individual ability to deliver its benefits correctly, safely, or reliably. Knowledge has both saved us and burdened us.*
>
> — Atul Gawande

Finance demonstrated that accountability structures shape how AI must be deployed; insurance extends that logic into a discipline where every decision is, by definition, a structured wager on uncertainty.

This chapter follows a single submission from arrival to bind. A mid-sized food manufacturer, an unusual occupancy mix, a fire at a sister site three years ago, a partial sprinkler retrofit covering the production hall but not the warehouse, a new owner with no prior insurance footprint with the carrier. None of this is in the rating engine's training set. None of it is in any single document, and the questions worth asking are not the ones on a checklist. By following the file across each stage of its life, the four operating models declare themselves in the order in which the work actually meets them, rather than in the order a textbook would list them.

That order is the point. An insurance organisation contains several kinds of cognitive work, and the mistake most insurers make is to treat AI as a single capability deployable uniformly across all of them. The submission's progress through the carrier's systems shows why that mistake is expensive: by the time the underwriter sits down to reason about it, the file has already been touched by three different operating models, and a fourth is waiting downstream for the wording.

20.1 Stage one: ingestion is invisible, and that is what makes it valuable

The submission arrives as a bundle. A broker email. A schedule of values in spreadsheet form. A loss run as a scanned PDF. Photographs of the partial retrofit, exchanged in the course of the placement. A surveyor's report from the prior carrier, attached without a covering note. None of this is structured for the underwriter; all of it has to be made structured before her time becomes worth spending on it.

This is Embedded Intelligence at full reach. Document ingestion, optical character recognition, extraction of structured data from broker submissions, classification of incoming correspondence and the production of routine reporting all run continuously and quietly inside the carrier's systems. The mark of well-designed embedded intelligence is that the underwriter never thinks about the mechanism: the submission arrives in her queue already structured, the schedule already parsed, the loss run already populated against the file. The technology is a property of the environment rather than a tool she picks up.

The discipline here is design-time. By the time the submission reaches the queue, the governance has either been done or not, and the consequences will surface unevenly across the workflows that depend on it.

20.2 Stage two: triage is where Autonomous Agency does the work the underwriter does not need to do

The same carrier handles thousands of submissions a week. Most of them are not the food manufacturer. Most of them are routine motor renewals, well-documented home risks and small commercial accounts whose criteria for placement are well defined and consistent application matters more than deliberation.

This is the proper domain of Autonomous Agency. Initial triage, routing of incoming notifications to the appropriate handler, first-touch assessment of low-value property risks and routine endorsement processing all benefit from agents operating within explicit objectives and guardrails. The criteria are codified. The variability is bounded. The volume rewards consistency that a human team would struggle to sustain across thousands of files.

The discipline is in the specification of the guardrails rather than in the cleverness of the model. The agent must know which matters to escalate, what tolerances apply to its decisions and how its outputs feed downstream reserves and reporting. Insurance carries reserve and reinsurance implications that automation must respect. An agent that under-reserves at scale creates a balance sheet problem before anyone notices.

> *A motor claims agent further along in the carrier's operation handles routine windscreen, minor collision, and theft notifications end to end. It verifies cover, applies the relevant excess, dispatches the assessor or repairer from the panel, sets an opening reserve from a calibrated table, and updates the file with each step.*
>
> *A handler sees the file only if the agent's confidence falls below threshold, the customer escalates, the reserve crosses an authority limit, or a sample is pulled for quality review. The agent is not pretending to be a senior adjuster, only doing exactly the work it has been authorised to do.*

Triage is what protects the underwriter's time. The food manufacturer submission has been held back from autonomous handling not because the agent is failing but because the agent's confidence threshold is doing exactly what it was specified to do: surfacing the unusual occupancy mix, the partial mitigation and the absent prior footprint as features the autonomous path was never authorised to clear. The submission lands on the underwriter's desk because the system upstream of her was correctly configured to recognise that it should.

Fraud screening sits in an interesting position between operating models. The detection of indicators across the inflow is well suited to autonomous operation. The determination of whether a particular claim is fraudulent remains a human responsibility, supported by partnership-mode investigation. The agent flags. The investigator reasons together with an AI partner about what the flag means.

The prize for getting this division of labour right is large. Deloitte's 2025 financial services predictions estimated that AI applied across the claims lifecycle could save US property and casualty insurers \$80 to \$160 billion in fraudulent claims by 2032, in a market where soft fraud, roughly 60 per cent of incidents, has detection rates as low as 20 to 40 per cent today (Deloitte 2025b). Those numbers are achievable only when agents do the screening they are good at and investigators, partnered with AI, do the reasoning that decides whether a flag becomes a finding.

20.3 Stage three: the underwriter's session is Paired Cognition, and the gain is reference breadth rather than speed

The underwriter has seen perhaps four genuinely comparable risks in twenty years. She knows the construction class. She knows the occupancy hazard, broadly. What she does not have in working memory at the moment of decision is the full set of treaty conditions that will apply, the regulatory positions her carrier has taken on partial mitigation, and the half-dozen technical bulletins issued since the last comparable risk crossed her desk. The questions she wants to ask are not "what is the answer," but "what am I missing." That is the question Paired Cognition is built to answer.

The value of partnership in underwriting is most often misdescribed as efficiency. Speed matters, but it is not the deepest benefit. The deepest benefit is reference breadth. Experienced underwriters have always struggled to keep treaty wordings, endorsement libraries, regulatory positions and accumulated technical bulletins simultaneously in view when reasoning about a single submission. The mind holds what it holds. The corpus of relevant material grows year on year. The slice of precedent that happens to be top of mind on any given Tuesday is necessarily narrower than the slice that ought to inform the decision.

In a Paired Cognition workflow, the underwriter no longer chooses between depth and breadth. The AI maintains breadth across the corpus while she applies depth of judgement to the specific case. The shared mental model that emerges through dialogue is richer than either party could construct alone, and the trail of reasoning is far more legible to peer review.

This matters in markets with active regulatory oversight. In the Australian context, the Australian Prudential Regulation Authority's expectations around governance, risk appetite, and decision traceability sit alongside the Australian Securities and Investments Commission's conduct obligations and the various state-based statutory schemes that shape liability cover. An AI partner that surfaces the relevant constraints in the moment of decision reduces the likelihood that a subtle obligation is overlooked. It also produces a record of reasoning that holds up under later scrutiny, which is its own form of value.

The specialty case sharpens the argument. A financial lines underwriter accustomed to professional indemnity for accountancy practices, asked to consider a submission from a mid-sized engineering consultancy, faces familiar base structure and unfamiliar exposures: different standards of care, different project documentation conventions, a different way that liability typically attaches, different relevant case law. With an AI partner holding the relevant material in active context, she can frame the questions she would have learnt to ask in five years on a specialty desk, in the first hour on this submission. She is still the underwriter. She is no longer pretending the second specialty is the first one with different paperwork.

The pattern generalises. Every insurer carries underwriters whose technical excellence in one class is disproportionately strong relative to their depth in adjacent classes. Paired Cognition makes the gap between strength and stretch a smaller barrier to good underwriting, without dissolving the gap altogether.

20.4 Stage four: drafting the endorsement is Task Delegation, and the boundary that defends it matters

Once the underwriter has made the call on the food manufacturer, the work shifts shape. The schedule needs amending. Endorsement language needs drafting against the accepted partial mitigation position. A short note to the broker needs writing. A first draft of the binder, with the carrier's standard conditions overlaid against the agreed variations, needs to be in the broker's inbox before close of business.

This is Task Delegation. The practitioner specifies the task; the AI produces a draft; the practitioner reviews and finalises. The cycle is fast and the outputs are predictable. Drafting tasks with clear specifications, such as endorsement language for a defined change, standard correspondence to insureds, policy summaries for distribution partners and first drafts of decline letters, are well suited to a delegate-and-review pattern.

The boundary worth defending is between drafting tasks where the practitioner can recognise quality on review and judgement tasks where review alone is not a sufficient quality control. A coverage opinion is not a drafting task in this sense, even though it produces a written output. The reasoning that produces it requires the sustained engagement of partnership rather than the delegate-and-review rhythm of task work. This is the directed-execution distinction developed in Chapter 11.

The risk in mature insurers is that Task Delegation becomes the default everywhere, including in places where it cannot carry the load. A coverage opinion written as a task can often read as a coverage opinion until the matter is litigated. The shape of the underlying reasoning, or its absence, only becomes visible under pressure.

20.5 The same shape recurs across claims, portfolio review and wordings work

Claims work mirrors underwriting in its structure. The simple, well-documented claim follows a defined path. The complex claim, particularly in liability and large property lines, demands sustained professional adjudication, and the same four-stage progression holds: ingestion, triage, paired adjudication, drafted output.

> *A bodily injury claim arrives with disputed liability, contested quantum, and potential subrogation against a third party. The medical evidence is incomplete. Two witnesses contradict each other on the sequence of events. The policy wording has an exclusion that may or may not apply, depending on how the activity at the time of the accident is characterised. The reserve carried on the file was set early, before the disputed elements emerged, and is now visibly under-pressure.*

Coverage analysis, causation, quantum, contributory factors and the interaction of primary and excess layers all converge in matters that resist tidy resolution. Because a change in one assumption ripples through the others, the work cannot be cleanly decomposed into independent steps. Paired Cognition supports it by holding the full evidentiary picture alongside the relevant coverage analysis and precedent. The adjuster interrogates the file with the AI as a thinking partner, testing alternative theories of causation, examining how comparable matters have settled, and pressure-testing reserve assumptions before they harden. Decisions remain the adjuster's. The picture against which they are made is more complete.

The scale of what is now achievable is no longer hypothetical. McKinsey's 2025 review of AI in insurance reported that Aviva, having deployed more than

80 AI models across its claims operation, had cut liability assessment time on complex cases by 23 days, improved claim routing accuracy by 30 per cent, reduced customer complaints by 65 per cent, and saved over £60 million in motor claims in 2024 alone (McKinsey & Company 2025b). These are figures the carrier and its consulting partner were both motivated to publish, and the headline numbers should be read with the usual caution that vendor-disclosed performance attracts; even discounted, however, they describe an operation that has reorganised genuine quantities of complex work around the framework rather than around tooling. The point is not that the models did the adjudication but that adjusters working alongside them, with the file properly framed and the precedent properly surfaced, were able to resolve disputed matters faster and more consistently than the unaided practice could sustain.

Portfolio review and wordings work follow the same logic across a longer timescale. A portfolio manager assessing concentration risk across a commercial book, or a wordings specialist reviewing the cumulative effect of endorsement drift across a renewal cohort, faces the same combination of contextual judgement and broad reference need. The AI compares wordings across a book, surfaces inconsistencies, identifies clauses that have aged poorly against shifting case law and proposes clarifications for the specialist to evaluate. The specialist retains authorship of the wording position. The work unfolds over hours or days, with the shared mental model deepening as the analysis progresses, and the sustained engagement is what makes Paired Cognition worth the investment. A wordings review compressed into a single transactional prompt produces a memo. A wordings review conducted as a partnership over a week produces a position the firm can defend.

The pattern carries through to reinsurance. Treaty placement, facultative submissions, and recovery analysis all involve dense documentation and intricate interdependencies, and a reinsurance specialist working with an AI partner can maintain a clearer view of how cessions, retentions and event definitions interact, particularly when treaty wordings have evolved across renewal cycles in ways that no single person has tracked end to end.

20.6 What the submission shows about a coherent insurance operation

Followed end to end, the food manufacturer file has touched every operating model the carrier runs, in the order the work required them. Embedded Intelligence structured the inputs. Autonomous Agency cleared everything around the file that did not require senior judgement and held this one back because it did. Paired Cognition was where the actual underwriting reasoning happened. Task Delegation produced the documents that left the carrier's systems.

The leadership task is to recognise this distribution clearly and resource it accordingly. Underinvestment in partnership tooling for the senior reasoning step, or overreach of automation into matters that require contextual judgement, both produce the same downstream symptom: deteriorating loss ratios paired with rising expense ratios. The combination is often suggestive, and worth investigating as a possible operating-model mismatch somewhere in the chain.

The framework is not a slogan. It is a way of allocating cognitive work to the model in which it actually performs.

The submission, in the end, is bound. The reasoning is on the file. The reinsurer is comfortable. The auditor, when she comes through next quarter, will see the trace of a decision made with everything in view rather than the trace of a decision made under the pressure of a Tuesday afternoon and the slice of precedent that happened to be top of mind. The treaty conditions were checked. The technical bulletins were surfaced. The partial sprinkler retrofit was weighed against the carrier's stated position on partial mitigation, not against whatever the underwriter happened to remember about it. The gain is not a faster underwriter but a better-supported one, working with the full weight of the corpus behind every decision that matters, on a file the system upstream of her was right to put in front of her in the first place.

Chapter 21

AI in Legal Practice: Research, Contracts and Strategy

> *Read not to contradict and confute; nor to believe and take for granted; nor to find talk and discourse; but to weigh and consider.*
>
> — Francis Bacon

Where insurance applies expertise to bound and price uncertainty, law applies it to interpret meaning: a shift from the quantitative to the textual that brings its own failure modes when AI enters the picture.

A senior litigator in Sydney files a submission citing three cases that do not exist. The cases are plausible. The citations follow the right format. The reasoning attributed to each is consistent with the surrounding argument. They were generated by an AI tool, included in the brief without verification, and discovered only when opposing counsel went to read them.

This is not a hypothetical. Variations of it have happened in courtrooms across multiple jurisdictions, including Australia, often enough that practice directions and disciplinary findings now exist on the topic. In 2025 a Victorian solicitor was stripped of principal-practice rights, placed on two years of supervised practice and barred from holding trust money after submitting AI-fabricated citations to the Federal Circuit and Family Court of Australia, the first Australian sanction of its kind. A separate Victorian King's Counsel apologised in the Supreme Court for AI-generated fake quotes in a murder matter the same year (Information Age 2025). It is also not, primarily, a story about AI. It is a story about a senior practitioner who treated AI output as finished work rather than as raw material, and who skipped the verification step that would have caught the problem.

That single failure mode shapes how legal AI adoption needs to be approached. The work is unusually well suited to partnership. It is also unusually unforgiving of partnership done badly.

21.1 Why is legal an unusually sharp version of the adoption question?

Legal work is heavily textual, pattern-rich, and built on cumulative reasoning across decades of authority. On the surface, that profile suggests a near-perfect fit for current AI capability. The same work, however, carries strict accuracy requirements, professional responsibility obligations, privilege constraints and confidentiality duties that make naive adoption genuinely dangerous.

Two failure patterns dominate in firms that have moved early. The first is timidity: confining AI to drafting routine letters and formatting documents, treating it purely as a delegation tool and missing the much larger partnership opportunities in research, analysis and strategy. The second is recklessness: leaning on AI output without adequate review, producing memoranda that contain fabricated citations, misapplied precedent or characterisations of holdings that do not survive contact with the source.

Both patterns trace back to the same missing piece. Neither approach distinguishes between kinds of legal work, and neither matches the work to the right operating model. A general-counsel team that treats every AI use case the same way will get every use case wrong in different ways.

The four operating models map onto legal practice cleanly once that distinction is made. Paired Cognition belongs in research, contract analysis and case strategy, where context, judgement and sustained reasoning dominate. Autonomous Agency suits discovery, due diligence and contract portfolio review at scale. Task Delegation handles well-specified drafting. Embedded Intelligence sits inside matter management, document classification and intake routing. The remainder of this chapter walks through each in turn, and then addresses the professional responsibility frame that constrains all of them.

21.2 Legal research is one of the strongest cases for Paired Cognition in any profession

Legal research is the clearest example of work suited to Paired Cognition anywhere in this book. The lawyer holds the matter context, the client objectives, the procedural posture and the strategic frame. The AI brings breadth across large bodies of authority, speed in surfacing adjacent material and the ability to draw out lines of argument a human researcher under time pressure might miss. Neither side replaces the other.

The session looks like a conversation, not a request and response.

> *A senior associate is preparing advice for a corporate client on whether a proposed restructure will trigger an obligation under a specific section of the Corporations Act. She begins by externalising what she already knows: the commercial shape of the transaction, the client's risk appetite, the previous advice given on adjacent points and the regulatory posture of ASIC on this kind of structure over the last three years.*
>
> *The AI proposes lines of inquiry: primary authorities, secondary commentary, recent enforcement activity and comparable structures considered in published rulings. The associate redirects when a line is unproductive, asks for the actual text of a holding when a summary feels thin and tests how a principle would apply to the specific facts in her matter. Two hours in, she has covered ground that would have taken a day alone, and the conclusions are firmer because the counter-arguments have been engaged rather than assumed away.*

What the AI cannot do is decide whether a precedent is genuinely on point in the specific matter. That assessment requires understanding the client's commercial position, the appetite for risk, the procedural posture and the way local courts treat particular lines of authority in practice. Those judgements stay with the lawyer. They cannot be delegated, even when the underlying retrieval and synthesis are materially accelerated.

The practical effect is that a lawyer working in partnership covers more ground in a research sprint than they could alone, and reaches firmer conclusions because alternative arguments have been tested rather than waved past. Coverage rises. Depth rises. Time spent on mechanical retrieval falls. The hours released go into the parts of the work that actually need a lawyer.

21.3 Hallucination, plainly named

Before turning to verification as a method, the failure mode it answers must be named plainly. Current legal AI systems hallucinate: they generate plausible-looking citations to cases that do not exist, paraphrase holdings into characterisations the source does not support, and confidently attribute reasoning to judgments that were never written. The frequency varies by tool, by prompt and by domain, but the failure mode is structural rather than incidental, and it does not vanish as the underlying models improve at the rate marketing materials sometimes suggest. Stanford RegLab's 2025 study by Magesh and colleagues, despite vendor claims of hallucination-free output, measured Lexis+ AI hallucinating on more than 17 per cent of queries and Thomson Reuters' Westlaw AI-Assisted Research on roughly 33 per cent, defined as either a factually wrong statement of law or a citation that did not actually support the claim (Magesh et al. 2025).

This is the specific risk that distinguishes legal AI adoption from adoption in most other professional domains. A fabricated number in a marketing brief is embarrassing. A fabricated authority in a court filing has, in reported cases, been treated as a duty-of-candour breach. The asymmetry between how convincing the output looks and how easily it can be wrong is the asymmetry the lawyer's method has to close.

There is no software setting that closes it. Verification, done by the lawyer, is what closes it.

21.4 Verification: woven through, not appended

In a Paired Cognition workflow, verification is not a final checkpoint. It is woven through the work itself, performed against primary sources, by the lawyer who will sign the document.

The discipline has four concrete components, and they are worth naming because vague exhortations to "check the AI's work" tend not to survive a busy Friday afternoon.

First, every cited authority is read in full or in the relevant passage before it is relied on. Not the AI's summary of the authority. The authority. If the case is unreported or the database returns no result, that is itself the answer: the citation does not exist or has been mis-rendered, and the proposition it was supposed to support is now unsupported.

Second, every characterisation of a holding is confirmed against the source text. The AI's account of what a case stands for is treated the way a partner treats a junior's headnote: useful as a pointer, not as evidence. The lawyer reads the passage that does the actual work and confirms that it carries the weight the argument is placing on it. The procedural posture is checked, because a holding from an interlocutory application carries different weight than a final judgment.

Third, every statutory provision quoted or paraphrased is pulled up in the consolidated version current as at the relevant date. Statutes amend. Drafts and commentary go out of date. The provision the matter turns on is read in its current form by the person responsible for the advice.

Fourth, the chain of reasoning that connects each authority to the proposition it supports is constructed by the lawyer, not borrowed from the AI's narration. If the lawyer cannot, from her own reading, walk through how each cited authority supports the conclusion, the work is not finished, regardless of what the document looks like.

> *A partner reviewing a junior solicitor's draft memorandum does not assume the cited cases say what the memo claims they say. She checks the ones the argument turns on. She reads the holding rather than the headnote. She confirms the procedural posture.*
>
> *The same discipline, applied to AI-generated material, is what makes the partnership safe. The lawyer who skips it has not adopted Paired Cognition but has adopted something else and called it by that name.*

The lawyer who treats AI output as a finished product, rather than as a working draft to be tested against primary sources, is operating outside the partnership model. They have collapsed Paired Cognition into a kind of trusting delegation that the technology does not support and that professional responsibility does not permit.

The diagnostic question is sharp and worth carrying into every brief. If a court asked the lawyer to walk through how each authority cited supports the proposition for which it is cited, could she do it from her own reading of the authority? If the answer is no, the work is not yet finished.

21.5 Contract analysis combines systematic coverage with judgement that does not transfer

Contract analysis is another domain where Paired Cognition produces strong results. AI systems can read a contract quickly, identify clauses that depart from market standard, flag missing protections and surface comparable language from precedent or from the firm's own playbooks. What they cannot do is decide whether a particular departure matters in this deal, for this client, at this stage of the negotiation.

Consider a clause that allocates risk in a way that is unusual but commercially defensible. The AI may flag it as a deviation. Whether to push back hard, accept the position with disclosure to the client or trade it against another point is a judgement call. That judgement depends on the client's commercial priorities, the leverage in the deal, the relationship with the counterparty and the lawyer's reading of where the negotiation is heading.

> *A senior solicitor is reviewing a heavily negotiated services agreement on a Friday afternoon. The AI flags eleven points worth attention.*
>
> *Three are genuine issues that the client would want to know about. Five are deviations from the firm's standard playbook that, in this commercial context, the lawyer is comfortable accepting. Two are matters where the playbook itself is now out of date, and the deviation is actually the better position. One is a genuine drafting error that the lawyer would have caught eventually but might have missed under time pressure. The list is the AI's contribution. The triage is the lawyer's, and the triage is most of the value.*

In a paired contract review, the AI provides systematic coverage and the lawyer provides the strategic and contextual reading. The output is faster than either alone could produce, and typically more thorough. Issues that a fatigued reviewer might miss in clause forty-seven of a long agreement get caught. The strategic shape of the deal stays under the lawyer's control.

This pattern extends naturally to negotiation support. A lawyer working through successive drafts can use the AI to track positions across versions, summarise the counterparty's typical drafting moves and propose fallback language, while making the actual negotiation decisions herself. The AI is in the room as a partner, not at the table as a negotiator.

21.6 Case strategy is the most demanding form of legal partnership

Case strategy development is the most demanding form of partnership in legal work. It involves sustained reasoning over weeks or months about how the law applies to the facts, how precedent should be framed, what arguments to run and which to abandon, how witnesses will perform and how to anticipate the other side's moves. It is the work that most distinguishes excellent litigators from competent ones.

AI contributes to this work as a thinking partner.

A litigator can stress-test a theory of the case by articulating it to the AI and asking for the strongest counter-arguments, not the obvious ones, but the ones a thoughtful opposing counsel might actually run. She can explore how an appellate court has historically treated a class of argument and where the current bench may be moving. She can map evidential gaps in a draft submission and think through what additional material is needed before a directions hearing. She can rehearse cross-examination strategies by externalising the witness's likely positions and probing for inconsistency.

The decisions about which arguments to run, how aggressively to pitch them, what to concede in exchange for what and how to read the tribunal remain with the lawyer and ultimately with counsel briefed on the matter. These are professional judgements that draw on experience, ethical obligations and a feel for tribunals that AI does not have. The partnership accelerates the thinking without replacing the thinker.

Done well, this kind of paired strategic work raises the quality of analysis available to the client. Arguments are tested earlier. Weaknesses are identified before they are exposed in court. The strategic frame is sharper because it has been challenged. The hours saved on mechanical work are reinvested in better thinking about the matters that actually need it.

21.7 High-volume document work belongs in Autonomous Agency

Not all legal work calls for partnership. Discovery review, due diligence document review and contract portfolio analysis at scale are different kinds of work. They involve applying defined criteria to large volumes of material, with the criteria themselves relatively stable across the exercise. This is Autonomous Agency territory.

In an Autonomous Agency configuration, the supervising lawyer or partner sets the objectives and the guardrails up front. What categories of document matter. What constitutes a privileged communication. What triggers escalation to human review. What confidence threshold the system must meet before classifying autonomously. The AI then works through the corpus, classifying, extracting and flagging exceptions. Humans review the exceptions and audit samples of the routine classifications.

> *A transaction team is conducting due diligence on the acquisition of a logistics business with eleven thousand customer contracts. The deal team needs to know which contracts contain change-of-control clauses, which have unusual termination rights and which carry pricing protections that survive the transaction.*
>
> *A senior associate, working with the AI, defines what each category looks like and trains the classification on a sample set. The AI then runs the full corpus, returning a structured output, a list of flagged exceptions for human review and a confidence score on each classification. The associate audits a stratified sample, attends to the exceptions and signs off the work. The substance of the review is human, even though the execution at scale is not.*

This model has been used in e-discovery for years with earlier generations of technology, and modern AI extends both the accuracy and the range of work it can handle. Due diligence on a transaction is a natural application. So is contract portfolio review when a regulatory change requires identifying affected agreements across a large book of business.

The professional responsibility position remains clear. The supervising lawyer is accountable for the work product, regardless of how much of it the system produced. What Autonomous Agency transfers is execution, while authority and accountability stay with the lawyer who set the parameters and reviews the exceptions.

21.8 Task Delegation has its place, but not as the default mode

Task Delegation has its place in legal practice, particularly for drafting work that is genuinely well-specified. Standard engagement letters with defined parameters. Routine correspondence. Formatting and proofing. Conversion of dictated notes into structured drafts. Decline letters to prospective clients on conflict grounds. These are tasks where the lawyer can specify the output clearly and review it efficiently.

The risk is treating this as the only operating model, which is what most firms do in their first year of adoption. A practice that uses AI exclusively for delegation, while leaving research and analysis untouched, captures a small fraction of the available value. The drafting savings are real but modest. The larger gains come from partnership in the analytical work, where coverage and depth improve substantially.

A balanced legal AI strategy uses Task Delegation for work that genuinely fits it, Paired Cognition for analytical and strategic work, Autonomous Agency for high-volume defined work and Embedded Intelligence for matter intake routing, document classification and conflict checking inside the firm's systems. The mix reflects the actual shape of the practice rather than a single default mode imposed across everything. The directed-execution distinction developed earlier is what separates a coverage opinion produced through partnership from a coverage opinion produced as a delegated task; the surface output can look similar, and the underlying reasoning rarely is.

21.9 Professional responsibility does not transfer to the system

The single most important principle in legal AI adoption is plain. Professional responsibility does not transfer to AI systems, and it does not transfer to the firm's IT department, the vendor or the model provider. The lawyer who signs the document is the responsible practitioner. Duties of competence, candour to the court, supervision of those assisting and confidentiality apply in full whether the work was drafted longhand, by a junior solicitor, by a contracted paralegal or with AI assistance.

The Australian Solicitors' Conduct Rules, and the equivalent barristers' conduct rules across the states, have not, to date, bent around new technology. The duty to the court is paramount and personal. The duty of competence requires the practitioner to understand the tools she uses, including their failure modes. The duty of candour means not relying on authorities the practitioner has not verified. The duty of supervision means actually reviewing AI output before it leaves the practitioner's hands, in the same way a partner reviews a junior's work.

None of these obligations are new. AI changes how they must be discharged in practice, not what they require.

Privilege and confidentiality require careful thought about what information can be shared with which AI systems. Client information shared with a public AI service may not retain the same legal professional privilege protections as information held within a firm's own systems or within a contracted enterprise instance. The choice of tool, the contractual terms with the provider, the data handling configuration and the deployment posture can all matter for privilege analysis. A firm's general counsel and its risk function need to be involved in those choices, not consulted after the fact.

Some firms have responded by moving cautiously, with sanctioned tools, defined use cases and mandatory training. Others have responded by issuing blanket prohibitions, which tends to drive use underground rather than prevent it. The firms that have integrated AI most successfully have done so by treating professional responsibility, privilege and confidentiality as design constraints from the outset, woven into the choice of tools, the configuration of access and the training of practitioners. Rather than being a separate workstream applied after the fact, compliance is part of the architecture.

21.10 The training of junior lawyers is a question that deserves an honest answer

A great deal of how junior lawyers have historically learned to think like lawyers came from doing the very work that AI now accelerates. Pulling cases, reading them closely, distilling holdings, drafting first-cut research notes and comparing the partner's marked-up version to the original. The retrieval was the apprenticeship, even when the retrieval was not the most valuable part of the output.

If junior lawyers spend less time on that retrieval, where does the apprenticeship go?

The glib answer is that AI frees juniors to do higher-value work sooner. They do not, in fact, automatically rise to higher-value work just because the lower-value work has been compressed. Judgement is built by reading the same fifty cases that a partner read at the same point in her career, then disagreeing with the partner about what they mean. Removing the reading does not produce the disagreement.

Firms that have thought carefully about this are designing training in which juniors still read the authorities, still draft the first version and still defend their reasoning, with AI used as a check against missed material rather than as a substitute for the work. The output is sometimes slower than partnership-mode AI use would produce. The training value is preserved. The trade-off is conscious.

The skills that compound are the skills that get practised. If close reading is preserved, it remains a senior skill. If it is not, the profession will discover, over a generation, what the consequences of that loss actually are. This is a question for the profession, the law schools and the admitting authorities together, and one of the genuinely open questions that legal AI adoption raises. What firms can do today is name it rather than ignore it, and resist the temptation to treat junior time saved as pure efficiency gain when at least some of that time was producing something the firm needs juniors to have produced.

21.11 What changes for a legal team when this is done well

When a legal team adopts Paired Cognition seriously, several things shift at once. Research coverage broadens because more authorities are surfaced and tested. Research speed increases because retrieval and synthesis accelerate. Analysis quality improves because counter-arguments get genuine engagement rather than being noted in passing and set aside. Client value rises because the time freed from mechanical work goes into better thinking about the client's actual problem. Thomson Reuters Institute's 2025 *Future of Professionals* survey of more than 2,275 legal, tax and risk professionals projected that AI would free roughly five hours of professional time per week within a year and close to 240 hours annually, worth about $19,000 per professional. Eighty per cent of law-firm respondents expected AI to transform their work or have a high impact (Thomson Reuters Institute 2025).

The shape of legal work itself begins to change. Junior lawyers spend less time on pure retrieval and more time, ideally, on the judgement-heavy work that builds expertise, provided the firm has resolved the apprenticeship question deliberately rather than by default. Senior lawyers can engage more deeply with strategy because the supporting analysis is faster and more thorough. Matters that previously could not be staffed economically become viable, and the firm's capacity to serve clients well expands.

The teams that achieve this share a small number of habits. They treat AI as a partner in the genuinely analytical work. They are disciplined about verification, against primary sources, every time. They understand which operating model fits which task and use the framework deliberately rather than defaulting to one mode for everything. They take professional responsibility seriously enough to build it into how the technology is used, rather than treating it as a constraint to be worked around.

The Sydney litigator with three fabricated citations did not lack a tool. He lacked a method. The method, which is not complicated, is also not optional.

CHAPTER 22

Clinical AI: Decision Support That Keeps Humans in Charge

You matter because you are you, and you matter to the end of your life.

— Cicely Saunders

Legal practice established that professional responsibility cannot be transferred to a tool regardless of how capable the tool becomes; clinical practice raises that principle to its highest stakes.

A registrar finishes a long ward round and sits down to work through the patient she is most worried about. The presentation is layered. There are overlapping conditions, two medications that pull in opposite directions, a guideline that updated three months ago and a family conversation tomorrow morning that she has not yet rehearsed. She has spent eleven years training for moments like this one. The decision she is about to make sits squarely on her registration, her professional standards and her conscience.

This is the work that the cognitive partnership has the most to offer and the least permission to displace. Few domains illustrate the stakes as sharply as clinical practice. The clinician carries responsibility that no algorithm can absorb. The reasoning she is doing is also exactly the kind of evidence-saturated, context-heavy professional thought where AI contributes genuine value when it is positioned correctly.

Healthcare organisations cannot wave the resulting tension away. AI systems can read more literature than any human, recognise patterns across vast case datasets and synthesise guidelines at speed. They cannot examine a patient. They cannot weigh competing values across a family system. They cannot hold a difficult conversation about prognosis. They cannot accept liability for a decision. The operating model framework introduced earlier in this book offers a way to resolve that tension cleanly, by matching the right model of AI involvement to the right kind of clinical work.

22.1 A concrete tension, not a philosophical one

Consider the registrar at her desk again. The patient in front of her presents with overlapping conditions, contraindicated medications, evolving evidence and a personal context that shapes what good care looks like for this person specifically. The clinician brings years of training, examination findings, the patient's preferences and a duty of care that cannot be transferred to a tool. None of those inputs can be reduced to text on a screen.

An AI system, properly used, can support that work in real ways. It can surface comparable case patterns. It can summarise the current evidence base. It can flag interactions and trace the implications of recent guideline updates. None of this displaces the clinician's role. All of it can extend her reach into a body of knowledge that has long since outgrown what any individual mind can hold in active recall.

The question is not whether AI belongs in clinical practice. The question is which model of involvement fits which task, and how organisations preserve the clinician's authority while still benefiting from the partnership. Those questions have answers. They differ from one part of the clinical workflow to another, and that variation is the point.

22.2 Paired Cognition fits the reasoning, not the rest of the work

Clinical decision support, treatment planning and diagnostic reasoning sit squarely in Paired Cognition. These are tasks of complex contextual judgement, where the relevant knowledge is broad, the patient is unique and the cost of being wrong is significant. The AI and the clinician work the problem together, with a shared mental model of what is being decided and why.

In treatment planning, the AI contributes breadth. It can synthesise current research, distil relevant clinical guidelines, lay out the comparative evidence for available options and surface case patterns that resemble the one in front of the clinician. This is reference work at a scale no individual practitioner can match.

The clinician contributes the judgement that matters.

A geriatrician reviewing a frail eighty-four-year-old with new heart failure, mild cognitive impairment and a recent fall is not looking for an algorithm. She is working out what good care looks like for this specific person, given the family who will support her at home, the goals the patient has stated about hospitalisation and the medication burden that is already at the edge of what the patient can manage.

The AI can hold the current heart failure guidelines, the deprescribing literature, the falls evidence and the comparative data on the available agents, all visible at once, all queryable. It cannot decide. It can only ensure the geriatrician decides with everything in view.

Diagnostic reasoning follows the same logic. AI pattern recognition across literature and historical case data can help a clinician consider differentials they might otherwise reach more slowly. It can test the strength of an emerging hypothesis against a wider reference set. It is an input to clinical assessment, not an attempt to replicate it.

The shared mental model in clinical Paired Cognition is built the same way it is built in any other professional domain. The clinician externalises what she is seeing, what she is worried about, what she has already ruled in or out and what the patient has said about how they want to live. The AI structures that input, surfaces relevant evidence and asks the questions that test the reasoning. The model is then maintained across the encounter, the ward round, the multidisciplinary meeting and the follow-up, not rebuilt from scratch each time the clinician opens the system.

This continuity is what distinguishes partnership from a series of isolated lookups. A clinician who briefs an AI from cold every time she has a question gets useful answers and a fragmented body of reasoning. A clinician working inside a sustained shared model gets reasoning that compounds across the case, with the AI carrying forward the constraints, preferences and prior decisions that shape what the next question even is.

22.3 Where does authority actually sit?

There is a phrase that gets used loosely around AI in healthcare: shared decision-making, or joint responsibility or partnership in care. The phrase is meant to describe the cognitive collaboration. It often gets read as something else, as a softening of who is actually accountable.

Clinical authority is non-negotiable, and not as a stylistic preference: the requirement reflects the structure of healthcare itself.

The shared responsibility in clinical AI is asymmetric, and the asymmetry is load-bearing. The clinician shares the cognitive load of working through the problem. She shares the benefit of broader reference, while the responsibility for the decision remains hers alone and cannot be shared on any terms. Decision-making responsibility, the central axis of the four-model framework, sits entirely with her. The AI is a participant in reasoning rather than a co-signatory on the outcome. This is augmented authority, not equal partnership, and the distinction is the one that has to be understood before any of the rest of the architecture works.

The reasons run together: legal, regulatory, ethical and clinical. Registered practitioners carry obligations under the Australian Health Practitioner Regulation Agency that bind them personally to the standard of care they provide. Clinical liability attaches to the decision-maker. Informed consent rests on the patient's understanding of a recommendation made by an accountable human professional. These are not bureaucratic overlays on top of clinical work. They are the mechanism by which healthcare manages the risks inherent in acting on imperfect information about complex biological and personal realities.

An AI system has no professional registration. It has no scope of practice. It has no fiduciary duty to a patient. It has no capacity to be held to account in the way a clinician can. These absences are structural, and they do not become less true as the underlying model becomes more capable. A more accurate model is still not a registered practitioner.

When the asymmetry is built into how systems are designed and how clinicians are trained to use them, the partnership works. When it blurs, problems arrive quickly: clinicians defer to outputs they have not interrogated, and organisations grow confused about where accountability sits. Regulators begin to suspect that the human in the loop has become a formality. None of those failure modes are theoretical, and none of them stay confined to a single bad case once the cultural drift is established. In Dratsch and colleagues' 2023 *Radiology* study (Dratsch et al. 2023), when a deliberately incorrect AI suggestion was shown alongside a mammogram, inexperienced radiologists' BI-RADS accuracy collapsed from around 80 per cent to under 20 per cent; even very experienced radiologists, with more than 15 years of practice, fell from 82 per cent to 45.5 per cent. Experience attenuated the bias but did not eliminate it.

The protective design move is to keep the partnership transparent at the moment the decision is recorded. The clinical record should show the reasoning, not only the result. Where AI input contributed to differential diagnosis, treatment selection or risk assessment, the input should be visible and the clinician's reasoning on top of it should be visible too. This is not a compliance overlay. It is the same documentation discipline that has always made clinical decisions defensible, extended to cover the additional source of information that the partnership introduces. A clinician who cannot, six months later, explain why she accepted or rejected an AI suggestion has not yet adopted the model. She has adopted something looser and called it the model.

22.4 Most of clinical work is not clinical reasoning

Outside the consulting room and the bedside, much of clinical work is not clinical reasoning. It is documentation. Coding. Record maintenance. Scheduling. Referral letters. Discharge summaries. Billing notes. Prior authorisation paperwork. Inbox triage. The countless small administrative artefacts that accumulate around every patient encounter, every shift and every admission.

This work consumes hours that clinicians would otherwise spend on patients. It is also where the most direct gains from AI are sitting in plain view. For most healthcare organisations, the administrative burden is often the binding constraint on clinical capacity well before any sophisticated decision-support question becomes the right one to ask.

Documentation is the largest piece. A consultation that lasts fifteen minutes can generate ten minutes of typing afterwards if the note has any complexity to it. Multiply that across a daily list and the hours add up to most of a working day per week. Ambient documentation systems, which listen to the consultation and structure a draft note against the practice's template, change the shape of that work. The clinician dictates and converses naturally. The system produces a structured output. The clinician reviews, edits and signs. What was thirty minutes of post-clinic typing becomes five minutes of review. A 2025 randomised trial published in *NEJM AI* (Schubel et al. 2025) assigned 238 outpatient physicians across 14 specialties to one of two ambient AI scribe systems or to usual care; the scribe arms cut time-in-note significantly relative to control, with both arms also showing improvements in clinician burnout, task load and work exhaustion.

Coding and billing have a similar profile. The work involves applying defined rules to a defined record, with the result subject to review. Embedded systems can propose item numbers from the consultation note, surface common errors before submission and learn the practice's patterns over time. The clinician retains authority over the final submission. The mechanical pattern-matching that previously occupied a clinician or a billing administrator gets handled in the background.

Referral letters and discharge summaries are well-suited to drafting under Task Delegation. The clinician supplies the structured inputs: the diagnosis, the relevant history, the management to date and the question being asked of the receiving clinician. The AI assembles the letter against the practice's template. The clinician reviews before signing. A discharge summary that previously required half an hour of consolidation across a long admission can be produced as a working draft in two minutes, leaving the clinician to focus on the parts that genuinely require her judgement: the medication reconciliation, the follow-up plan and the explicit safety-netting advice for the GP.

Inbox management belongs in the same territory. Results notifications, repeat-prescription requests and simple administrative queries: a well-configured system can triage, draft routine responses for review and surface only what genuinely requires clinical attention. The clinician spends her time on the messages that matter, not on sorting them from those that do not.

These workflows fit Task Delegation and Embedded Intelligence, not Paired Cognition. Embedded Intelligence describes the deeper integration: AI capabilities sit inside the electronic health record, the practice management system or the coding workflow, where the system handles routine pattern matching and data structuring as part of its normal operation. The clinician interacts with a smarter tool rather than commissioning each piece of work explicitly.

> *A general practice running 15-minute consultations has a registered nurse, two GPs and a practice manager. The practice introduces ambient documentation that listens to the consultation, structures the note against the template, drafts*

the referral letter and prepares the billing code for review. The GP reviews each output before it is signed.

Over a working week, each clinician recovers roughly six hours that previously went to typing. None of those six hours produced a single clinical decision. All of them now go to patients, to teaching the registrar or to the kind of conversation the practice manager has been asking the GPs to have for months.

The justification is the same as in any other industry where Task Delegation and Embedded Intelligence fit. The work is well-defined. The outputs are checkable. The risks are bounded by review. The time recovered can be redirected to the parts of practice where human judgement is irreplaceable.

This is where the most concrete clinical capacity gain sits. Reducing administrative burden is one of the most direct contributions AI can make to a healthcare system under pressure, and arguably the contribution that matters most for the clinicians actually doing the work. It is also the area where the operating model selection is least contentious: the decisions delegated are administrative, the human review is fast and the gain is visible inside a fortnight. Organisations that begin AI adoption here, before reaching for decision support, typically end up with both a stronger foundation and a more receptive clinical workforce.

22.5 Patient communication needs a clearly drawn line

Patient communication sits in a more nuanced position than either the consulting room or the back office. Routine communications can be drafted under Task Delegation. Appointment reminders. Preparation instructions. Post-procedure follow-up messages. Standard educational material. The AI prepares structured content. The clinician or appropriately delegated staff member reviews and approves before anything reaches the patient.

The boundary that matters is between communications that convey information and communications that constitute clinical advice.

A reminder to fast before a procedure is informational. A response to a patient's question about whether they should change their medication is clinical. The two cannot be conflated, regardless of how convenient automation might seem.

Holding this line protects patients from the failure mode where automated communications drift from logistics into substantive clinical guidance without the safeguards that govern clinical advice in any other channel. The principle is simple. AI drafts. A clinician decides what counts as advice. Patients receive responses that come from accountable practitioners.

Done well, this expands a practice's communicative capacity without compromising care. Patients receive timely, clear, well-structured messages. Clinicians spend their attention where it matters. The partnership model holds, with the AI taking on volume and the clinician taking on judgement.

22.6 The Australian regulatory context defines the perimeter

Clinical AI in Australia operates inside a specific set of constraints that healthcare leaders need to internalise rather than work around. Patient safety is the first principle. It is codified through clinical governance frameworks that every health service is expected to maintain. Any AI introduced into a clinical workflow falls within those frameworks and must be evaluated accordingly.

Regulatory oversight comes through several channels. The Therapeutic Goods Administration regulates software that meets the definition of a medical device. Software intended to inform or drive clinical decisions, including many forms of clinical decision support, can fall within scope as Software as a Medical Device, with classification depending on the seriousness of the condition and the role the software plays in the decision. The TGA's reforms in this area have moved progressively closer to the international consensus, and the prudent approach for healthcare leaders is to assess decision-support tools for whether they require inclusion in the Australian Register of Therapeutic Goods before clinical use, not after. By the end of 2025 the United States Food and Drug Administration had authorised 1,451 AI- or machine-learning-enabled medical devices in cumulative total, with 295 cleared in calendar year 2025 alone: roughly 76 per cent of them in radiology, 9 per cent in cardiovascular medicine (US Food and Drug Administration 2025). Clinical AI is no longer rhetorical.

The Australian Health Practitioner Regulation Agency administers the registration of practitioners and, with the relevant National Boards, the professional standards that bind them. Those standards apply regardless of the tools the practitioner uses. The Privacy Act and associated state and territory health records legislation govern how patient information may be handled, stored and disclosed, including when it leaves a clinical system to be processed by an AI tool. None of these regimes treats the introduction of AI as a special exemption from the obligations that already apply.

These constraints have practical implications for how AI is procured, deployed and audited. Vendor claims must be evaluated against regulatory categories rather than marketing language. Data flows must be mapped and justified. Practitioners must be trained to use systems within their scope and to recognise the limits of any AI input. Clinical governance structures must extend to cover AI-assisted decisions just as they cover any other element of practice.

Organisations that treat these constraints as the starting point, rather than as obstacles, generally find that the operating model framework aligns naturally with what good clinical governance already requires. Authority sits where the regulators expect it. Augmentation is documented. The patient's interest remains primary.

22.7 Evidence-based practice gets a tool it has been waiting for

For clinicians committed to evidence-based practice, the partnership model offers something specific and overdue. The evidence base has grown beyond the point where any individual practitioner can keep current across the full breadth of even a single specialty. Guidelines update. Trials report. Comparative effectiveness data accumulates. The gap between what is known and what any one clinician can have at hand widens steadily, regardless of how diligently the clinician reads.

AI used as a reference tool narrows that gap. It can help a clinician orient quickly to the current state of evidence on a question. It can surface relevant guidance from authoritative sources. It can structure complex case considerations for review. Although none of this outsources thinking, it extends the clinician's reach into the evidence that should already be informing her judgement.

This is the augmentation framing applied to its proper subject. AI does not make a clinician more confident in decisions she has not earned. It makes the evidence base more accessible to clinicians whose judgement remains the point of the entire exercise. The partnership is meaningful precisely because the human contribution is irreducible.

Clinical leaders setting strategy for AI in their organisations have a clearer task once this framing is in place. Choose the right operating model for each kind of work. Preserve clinical authority where the regulations and the ethics require it. Use Task Delegation and Embedded Intelligence to lift administrative burden. Use Paired Cognition to extend clinical reach while keeping humans in charge.

The registrar at her desk does not need an AI that decides. She needs an AI that ensures she is deciding with everything in view, and that gives her back the hour she would otherwise have spent typing the discharge summary. The cognitive partnership, applied cleanly to clinical work, delivers exactly that. Nothing more. Nothing less. And in healthcare, that distinction is the one that matters.

The next chapter moves the framework to a sector with a different relationship to language and the public record. Clinical work places the human contribution closest to the relational core of the work and pays for misjudgement in welfare. Marketing and communications place their human contribution closest to creative judgement and institutional voice, and pay for misjudgement in attention, revenue and credibility. The same four operating models apply, but they distribute differently across work that is structurally split between analytical optimisation, creative argument and the testimonial line that ends up on the public record. The framework is the same. The discipline of choosing where each model belongs is a different exercise.

Chapter 23

AI in Marketing and Communications: Two Registers, One Discipline

> *Facts need testimony to be remembered and trustworthy witnesses to be established in order to find a secure dwelling place in the domain of human affairs.*
>
> — Hannah Arendt

A communications director sits down on a Sunday evening to draft a statement that will go out at seven the next morning. The story has been gathering for forty-eight hours. The chief executive is on a flight. The legal team has views that conflict with the chair's. Three reporters have already filed, and a fourth is waiting on a quote. The director has perhaps an hour to produce something that will hold the line for a day, after which everything becomes recoverable.

This is not a content production problem. It is a thinking problem under time pressure, with stakeholders the director can name and a political weather she has been reading for years. Capable language models now touch almost every layer of the work that surrounds this moment. The moment itself remains stubbornly human, and the seam between those two facts is where the operating model question for this part of the book gets answered.

Marketing and communications are usually treated as adjacent functions, and in many organisations they share a leadership line. They are not, however, the same cognitive job. Advertising asks how an organisation reaches an audience and is judged by whether the message lands. Communications asks how the organisation is believed once it has, and is judged by whether the institution remains credible the morning after. The first register is outward-facing and creative-commercial. The second is institutional and testimonial, closer in posture to legal drafting than to copywriting. Practitioners who treat these two registers as the same job tend to misapply AI in both directions: producing communications that read as marketing copy, or producing advertising that hedges itself into invisibility. The four operating models map across the combined territory cleanly enough once that distinction is in view.

23.1 The work splits across all four operating models, not one

Most functions covered in this book cluster around a single cognitive shape. A bank's credit function is dominated by analytical reasoning. A clinical practice is dominated by judgement under uncertainty. Marketing and communications are structurally split, and the split is not a problem to be resolved. Strategy, audience analysis, creative iteration, message development under regulatory pressure, performance management and programmatic distribution each pull toward a different operating model. A team that recognises this can structure its work accordingly. A team that does not will end up applying one model, usually Task Delegation, across a value chain that needs all four.

What follows walks the four models once for the combined domain, with examples drawn from both registers. The voice question is where the two registers most clearly separate, and it is the chapter's distinctive contribution.

23.2 Strategy and stakeholder analysis are Paired Cognition

Strategy is where the analytical and the interpretive meet. An advertising strategist building a brief is reasoning about brand position, competitive context, category dynamics, audience proposition and business objective at the same time. A communications strategist building a content architecture is reasoning about what the organisation will say across channels and over time, how messages reinforce one another, how a narrative arc plays across owned, earned and paid channels. Both are sustained, contextual reasoning rather than a sequence of discrete outputs.

This is Paired Cognition territory. Not because AI replaces the strategist's judgement, but because the analytical surfaces of strategic work benefit substantially from a thinking partner that can hold large bodies of context without losing the thread of the argument. The strategist externalises the brand, the category, the competitive set, the political environment and the brief in progress, then uses the partnership to test framings, surface tensions and pressure-test the logic of a proposition before it leaves the room. The strategist remains the author. The AI does not write the brief.

What makes this Paired Cognition rather than Task Delegation is the absence of a clean specification at the start. Strategy emerges through the dialogue. The practitioner cannot brief the work fully in advance because part of the work is figuring out what the brief should be. Trying to delegate strategy as if it were a deliverable produces a recognisable failure mode: a serviceable but generic plan that any organisation in the sector could have written.

Audience analysis sits in the same zone, with a cleaner division of contribution. Modern campaigns and content strategies draw on first-party data, syndicated research, social listening, search behaviour, panel studies and qualitative work. No human practitioner can hold all of that in mind at once. AI can ingest and structure it at a scale no individual planner could match, then surface the patterns. The strategist does what only a strategist can do, which is decide what it means for the brand and the institution.

Stakeholder analysis on the communications side sharpens the same point. The model can ingest organisational charts, public statements, regulatory filings, media coverage and legislative records. It can map relationships, surface inconsistencies and identify who has spoken about what, when and to whom. What it cannot do is read the room. It does not know which board member is quietly preparing to retire, which regulator is frustrated with which department, or which journalist owes whom a favour. This information lives in the heads of practitioners and the relational fabric of their networks. Models learn from the public record, and the public record is by design what organisations have decided to put there. Everything that explains the public record sits in private memory and relational trust.

> *A senior corporate affairs lead is preparing for a regulatory submission that will be politically sensitive. The model assembles a stakeholder map across the regulator, the relevant ministerial office, three peak bodies, the affected industry's largest competitor, two consumer advocacy groups and the financial press. It pulls every public statement made by named individuals across the past three years, identifies positions taken, flags inconsistencies and ranks likely receptiveness based on the public record.*
>
> *The lead reads it in twenty minutes. She then spends an hour with the model adjusting it. The deputy commissioner is more open than the public record suggests because she chaired a working group with the lead's predecessor.*

The peak body chair has a private grievance with the competitor that will shape his framing of any joint position. The consumer advocate quoted most recently is on parental leave and the file has shifted to a colleague the lead has never met. None of this is in the dataset. All of it changes the map.

The map is now usable, and it is a joint artefact. Treating the same work as Task Delegation, with the model producing a stakeholder map and the human merely reviewing it, misses the point. The map without the context is a list of names with quotes attached.

23.3 Creative iteration and routine production are Task Delegation

Creative development is where the operating model question becomes most charged, and most worth being precise about. AI can generate alternatives at a pace no creative team can match. Route after route, headline after headline, visual treatment after visual treatment. Whether this is useful, or whether it produces noise that drowns out the work, depends entirely on how the creative process is structured around it.

The productive frame is that AI accelerates iteration without replacing judgement. A creative team can use AI to expand the space of possibilities at the divergent phase, generating wider option sets earlier than the team could alone, then converge using the same disciplines that have always defined good creative work. The judgement about which route is right, which line lands, which image carries the brand stays with the creative director. For most production phases this is closer to Task Delegation than to Paired Cognition. The team has the idea. They direct AI to produce variations within a defined territory. They review what comes back, accept what works, reject what does not.

LIONS Advisory's 2025 *State of Creativity* report (LIONS Advisory 2025) observed that despite AI's pervasive influence across submitted work, the year's Creative Effectiveness Lions showed notably less evidence of AI directly contributing to measurable effectiveness, and concluded that "as AI tools become universally available, human judgment becomes more valuable, not less."

The same operating model fits a substantial volume of communications production. Newsletters, social posts, internal updates, partner communications, event collateral, routine media materials. Once the brief, voice and constraints are clear, the work is execution. USC Annenberg's 2025 *Global Communication Report* (USC Annenberg Center for Public Relations 2025), surveying more than a thousand communication professionals worldwide, found PR practitioners using AI most often for social media content (43 per cent), research and analytics (36 per cent) and press materials (36 per cent). Exactly the production layer this section describes.

For higher volumes and more standardised content, Autonomous Agency becomes viable. A model operating within clear objectives and guardrails can produce, schedule and adapt content across a campaign with periodic human oversight rather than per-piece review. The risk in this layer is over-extension. Content that looks routine often carries strategic weight, and a model operating autonomously will not notice when a message needs to shift because the political context has shifted. A weekly customer newsletter is fine until the week of an outage. A staff update is fine until the week of a redundancy round. Keeping autonomous production inside genuinely well-defined boundaries, and escalating to human attention when conditions move outside them, is the discipline.

23.4 Performance management is Autonomous Agency

Once a campaign is live, the operating model shifts again. Performance management is fundamentally about monitoring large volumes of data against defined objectives and acting within defined parameters. This is Autonomous Agency in the textbook sense.

AI can monitor campaign performance continuously, surface anomalies, optimise bids and placements within set guardrails, rebalance budgets across creative variants and alert the human team when something falls outside expected behaviour. The work runs at a tempo and a volume that does not benefit from a human in the loop on every decision, and the decisions themselves fit cleanly inside an objective and a constraint set. The discipline that makes this work is the same as in any application of Autonomous Agency. The objectives must be explicit. The guardrails must be defined: budget caps, frequency limits, brand-safety constraints, creative rotation rules, the conditions under which the system should escalate rather than act. The control point is the configuration, not each individual decision.

The pattern recurs across performance disciplines. Attribution, mix modelling, audience-level reporting and incrementality analysis all benefit from AI that can process the data at scale and structure findings for human interpretation. The decisions about what to do with those findings remain with the strategist and the client. Autonomy at the execution layer; partnership at the interpretation layer.

23.5 Distribution and monitoring are Embedded Intelligence

The IAB and PwC's 2025 *Internet Advertising Revenue Report* (Interactive Advertising Bureau and PwC 2026) recorded US digital ad revenue of $294.6 billion, with programmatic alone reaching $162.4 billion, up 20.5 per cent year on year, and the report explicitly framed automated buying as "laying the groundwork for agentic AI-driven media buying". A pipeline at that scale does not behave like a tool. It behaves like infrastructure.

Programmatic media buying is the clearest example in advertising of Embedded Intelligence at full reach. The decisions about which impression to bid on, at what price, in what context, against which audience signal happen inside automated systems running at speeds and volumes that no human team participates in directly. The human role is upstream: strategists and planners define the audience, the objectives, the brand-safety parameters and the channel mix; trading teams configure the systems and review their outputs.

Communications has its own embedded layer. Publishing across channels, monitoring coverage and conversation, tracking engagement, generating reports and flagging emerging issues. AI runs inside scheduling tools, social listening platforms, analytics dashboards and reporting pipelines. The practitioner may never interact with the model directly, but its presence is what allows a small team to monitor a complex information environment and respond when something matters. The operating model question here is whether the embedded systems are tuned to surface what actually requires human attention. A monitoring pipeline that flags every mention is useless. A pipeline that flags the three mentions that signal an emerging narrative is the difference between getting ahead of a story and chasing it. Designing those thresholds is itself a strategic act.

This is also the layer most easily mistaken for transformation. Buying a social listening platform that uses AI is not adopting AI in any meaningful sense. The embedded layer is real and valuable. It is also the lowest-leverage place to start if the practitioner above it is still trying to draft a holding statement from scratch on a Sunday evening.

23.6 Voice is where the two registers separate

The four-model walk gets the work into the right shapes. It does not yet describe what makes marketing and communications distinctive among the industries in this part. The distinctive contribution is the voice question, and it is here that the two registers most clearly separate.

Advertising voice is a creative-commercial property. A brand voice is not a style. It is an accumulated record of decisions a brand has made about itself, often over decades: what it will and will not say, how it sounds when it is being serious, what it sounds like when it is being playful, the line between dry and warm that separates this brand from every other brand in its category. A model trained on the brand's archive can mimic surface features and be steered, with worked examples, toward outputs that read as plausibly in-voice. What it cannot do is decide whether a particular line is in the voice or merely close enough to pass for it, because that judgement is not pattern-matching against the past. It is judgement about whether the line still sounds like the brand a year from now, in a market that will have shifted, against competitors that will have moved.

> *A senior copywriter on a long-running brand has spent years calibrating the line between dry and warm, the line that makes the brand sound like itself rather than like its category. A new campaign needs 40 headline variants across six product propositions. The copywriter writes three exemplars of the voice in this campaign's specific register, hands the task to AI with the brief, the brand book and the exemplars, and reviews the returned set.*
>
> *About a third are usable, about a third are nearly there and about a third miss the voice entirely. The copywriter selects, refines and rejects. The volume is real. The voice is preserved because the copywriter is the one judging.*

Communications voice is testimonial, and the cognitive job is different. Three differentiations carry the weight.

The first is institutional testimonial. When a chief executive issues a statement, the words become part of the record by which the institution will be held to account: by regulators, by employees, by journalists, by the courts in some matters and by the executive's own successors. The line that performs well in the moment is the wrong test. The relevant test is whether the institution can defend the line in six months, against a different context, possibly under hostile cross-examination. That is a judgement about evidentiary posture rather than about whether the sentence reads well. It is also why senior communications professionals so often sound like they are arguing against the line that everyone in the room thinks is good. They are reading it forward into the conditions under which it will later be read back.

The second is regulatory pressure. Communications work for regulated organisations, including banks, insurers, healthcare providers, listed companies and government bodies, operates under disclosure and conduct rules that have nothing to do with how the message lands. A line about future performance can be inadvertent forward-looking guidance. A staff message during an investigation can be an admission. A board chair's comment to a journalist can move a share price within minutes of being attributed. The cognitive work is not to produce a message that performs; it is to anticipate how every line will read against the rule set the organisation operates inside, and to know which rule set applies before the model produces the tenth elegant variation.

The third is internal compounding, and it is the differentiation most easily missed. A communications team supports many voices at once: the chief executive on the financial story, the chair on the political story, the chief people officer on the staff story, the head of public affairs on the regulatory story. Each voice carries an accumulated public position, sometimes built over years, sometimes reaching back to statements made under previous strategies that the current leadership has not formally retracted. Voice drift across that body of statements does not show up inside any one piece of output. It shows up across a year, when a regulator's submission letter quotes the chief executive against herself, or a journalist runs a story on the gap between what the chair said in March and what was said in October. AI-drafted material accelerates this risk because each individual draft passes review while the cumulative drift remains invisible.

Edelman's 2025 *Trust Barometer* and its November flash poll on AI (Edelman 2025a, 2025b) found that only 44 per cent of people globally were comfortable with businesses using AI; trust in AI sat at 32 per cent in the United States, 36 per cent in the United Kingdom and 39 per cent in Germany. "Someone like me" was about twice as trusted as a CEO to tell the truth about AI. Peer voice beats top-down corporate voice. The trust environment that surrounds AI-drafted communications is not neutral, and the institutional voice does not become more credible by being produced more efficiently.

23.7 Crisis communication is where the framework earns its place

Crisis communication compresses the entire discipline into hours or minutes. A practitioner must understand what has happened, anticipate how it will be perceived, identify which stakeholders need to hear what and in what order, draft holding statements, brief spokespeople and coordinate across legal, executive and operational teams. The reasoning is complex, the stakes are material, and the clock is unforgiving.

This is the clearest case for Paired Cognition anywhere in the combined domain. Task Delegation fails in a crisis because the brief is moving faster than the briefing. Autonomous Agency fails because the cost of a wrong public position is not recoverable inside the campaign. Embedded Intelligence is, at most, the monitoring layer that flagged the issue in the first place. What the situation calls for is sustained reasoning between a practitioner and a model across a window of acute uncertainty, with the framing being rewritten as the picture changes.

> *A consumer brand is dealing with a product safety issue late on a Friday. Initial reports are partial and the technical picture is still moving. The communications director runs a Paired Cognition session and externalises everything she knows, every uncertainty, every constraint. The model holds the picture as it grows. It drafts a holding statement, a longer customer email, talking points for the chief executive's call to the regulator and a set of internal messages for staff who will face questions over the weekend.*
>
> *As legal sends back amendments, the model integrates them across all four artefacts and flags an inconsistency between the customer email and the staff brief. As the technical picture sharpens, the director adjusts the framing once and the suite updates together. The artefacts that go out at midnight are tighter*

than the team would have produced in twice the time, and they hang together because they were drafted as a single coherent set.

The model holds parallel artefacts coherent under shifting facts. The director keeps her attention on the calls only she can make: what is actually true at this hour, what can be said now versus later, what risks legal exposure, what protects relationships, what reflects the organisation's values. None of that is delegable. All of it depends on context the model does not have and consequences the model does not bear.

23.8 What is preserved

The four operating models do not tell marketing and communications professionals what to do. They offer a vocabulary for deciding where the practitioner's judgement is the differentiator and where it is overhead. The work that defines the combined function sits in Paired Cognition: strategy, audience and stakeholder analysis, message development for high-stakes situations, crisis response and internal change. The work that fills the day sits across Task Delegation, Autonomous Agency and Embedded Intelligence: production drafting, scheduled content, performance optimisation, programmatic distribution, monitoring and reporting.

A function that gets this distribution right looks different from one that does not. Senior practitioners spend more of their week on the work that requires their judgement, because the work that does not has been pushed cleanly into the modes where it belongs. Junior practitioners learn faster, because they see senior practitioners reasoning aloud with a model and absorb how the reasoning is done. Strategists spend less time assembling material and more time interpreting it. Creatives spend less time producing variations and more time choosing among them. The mechanical layer of the job, the parts that consume hours and produce no insight, gets compressed. The judgement layer expands into the room that compression creates.

The Sunday evening statement still has to be written by someone who knows the chair, the chief executive, the reporter and the sector. The 40-headline brief still has to be selected and shaped by a copywriter who can hear when a line sounds like the brand and when it does not. Those things have not changed, and they will not. What has changed is what surrounds those moments: the speed at which the supporting work can happen, the breadth of reference the practitioner can bring into the room, the volume of routine output the function can sustain without burning out its best people.

The voice on the public record still belongs to the institution, and to the people whose names will be on it when it is read back. That remains the work, regardless of how good the model gets at drafting.

Chapter 24

AI in Technology Teams: Architecture, Product and Operations

Always design a thing by considering it in its next larger context — a chair in a room, a room in a house, a house in an environment, an environment in a city plan.

— Eliel Saarinen

Marketing and communications professionals shape how organisations speak to the world; the technology teams in this chapter build the systems through which that world increasingly runs.

Engineering organisations were the first to adopt AI seriously, and many of them have stopped advancing because of it. They wrote the early tutorials, ran the first internal hackathons and produced the case studies the rest of the enterprise reads. They normalised the assistant in the editor, the autocomplete in the terminal and the agent in the pull request. Then, having moved further than any other function, a large share of them quietly converged on a single mode of working with AI.

That mode is code generation under Task Delegation. A developer prompts an assistant, accepts or rejects suggestions, and moves on. The tooling for this is excellent and getting better. The pattern is so visible that it has come to stand, in many engineering cultures, for the whole of what AI in technology is.

It is not.

What the default leaves on the table is more interesting than what it picks up. A function that uses AI only to write code faster is improving its execution speed on whatever it has decided to build. It is not improving the quality of the deciding. The architectural choices, the product trade-offs, the cross-system reasoning that determines whether the next year of engineering effort is well spent: none of this is what code generation accelerates. The decisive axis, as in every other industry, is decision-making responsibility. In technology that axis runs across architectural reasoning, product discovery, code production, autonomous task execution and continuous operations, and each of these calls for a different distribution of responsibility between human and machine. Engineering teams that map their work to the four operating models, rather than defaulting to one, gain leverage on the higher-value problems their organisations actually face.

24.1 Architecture is the clearest case for sustained partnership

Architecture and high-level system design are the work where Paired Cognition produces the most leverage in a technology organisation. They are also where it is most under-used, and the gap between the two is where the function quietly forfeits most of the value AI is now capable of providing.

The argument for sustained partnership in architecture rests on the structure of the decisions themselves. An architect choosing between an event-driven decomposition and a synchronous service mesh is not executing a discrete task with a known answer. The decision depends on team topology, latency budgets, regulatory constraints, the maturity of the platform tooling, the migration story for whichever legacy systems are in scope and the political weight of the vendor relationship currently in favour. None of these can be reduced to a specification. They shift as the architect learns more about the domain. They interact with each other in ways that resist clean separation, and the interactions are themselves part of what the architect is being paid to see.

This is sustained reasoning across competing constraints. It is iterative, contextual and judgement-laden in ways that make Task Delegation a category mistake. There is no brief that captures the work, because the brief is part of what the work produces.

The trade-offs compound. A choice that looks correct against latency and throughput may quietly increase operational burden on a team that cannot absorb it. A pattern that fits one regulatory regime may make a future expansion into another regime materially harder. A vendor decision that simplifies a current migration may concentrate platform risk in ways that only become visible during the next outage. None of these second-order effects appear when an architect treats the model as a search engine and asks discrete questions. They appear when the architect and the model hold the constraints in shared working context and stress-test the design against them across an extended dialogue.

In this mode, AI contributes technical breadth and pattern analysis at a scale no individual architect can match. A modern model has read more architectural decision records, more post-mortems and more vendor whitepapers than any one person could absorb across a career. It can surface analogous designs from adjacent industries, point at failure modes that have appeared in similar topologies and stress-test assumptions by playing scenarios forward. What it cannot supply is the organisational memory of why a previous CTO mandated a particular database, or which platform team will refuse to operate a service they did not build. The architect provides that context. The architect integrates the competing concerns. The architect is accountable for the choice.

The shared mental model emerges through dialogue, and the dialogue is where the value sits.

> *A senior architect at a fintech firm working through a payments rearchitecture spends the afternoon in this mode. She begins by sketching the constraints out loud: regulatory reporting timelines, the existing reconciliation pipeline, the cost of a card-scheme certification cycle, the team's familiarity with the candidate event broker.*
>
> *The model holds these in working context and challenges the sketch. It asks what she expects the failure mode to be when the broker partitions during a settlement window. It reminds her that one of the constraints she stated implies a stronger consistency guarantee than the design currently provides. By the time she writes the decision record, the trade-offs are explicit and defensible, not because the model authored them but because the model would not let her leave them implicit.*

Architects who treat AI as a search engine miss this entirely. They ask it discrete questions and paste the answers into a document. The work that ought to be Paired Cognition collapses back into Task Delegation, and the architect ends up with a faster way to produce the same shallow design they would have produced anyway. The output looks the same. The thinking behind it has been quietly hollowed out.

24.2 Product and design work belong in the same partnership

Product work sits in the same territory, and for the same reasons.

Requirements reasoning, prioritisation analysis and design trade-offs are not tasks with crisp outputs. A product leader weighing whether to invest in a new onboarding flow or a payments hardening initiative must reason simultaneously about cohort behaviour, regulatory pressure, the sales pipeline, accumulated technical debt and the morale of the teams who would do the work. The framing of the question evolves through the conversation. The product leader begins one way, the model surfaces a consideration that reframes the problem, and the leader brings in new information that further reshapes it. The output is not a document but a sharper understanding, which then informs whatever artefact comes next.

Discovery follows the same pattern. The questions are layered and the threads have to be held simultaneously.

> *A product manager at a healthtech company is exploring whether to build a clinician-facing decision support feature. The relevant questions are not in series. What is the regulatory pathway in each target market, and how does it differ for a feature that recommends versus one that informs? Which clinical workflows would the feature interrupt, and at which decision points would clinicians most plausibly trust it? What evidence would they require before relying on an output they did not generate themselves? How does any of that interact with the existing integration roadmap?*
>
> *The product manager works through these in dialogue with the model across an afternoon. The model holds the threads. The product manager moves between them without losing the connection between regulatory pathway and clinical adoption, or between adoption and integration cost. The output is a sharpened brief, not an answer.*

Design trade-offs at the engineering boundary respond to the same approach. A senior designer working with engineering on the shape of a new SDK can use AI to play out how different API designs would feel inside customer code, how they would version and how they would interact with the existing surface area. The designer remains accountable for the choice. The model accelerates the exploration and surfaces costs that would otherwise emerge only after release.

This entire band of work, architecture and product reasoning and design trade-offs together, is where engineering organisations leave the most value on the table. It is also where the public discourse about AI in technology has the least to say, because the artefacts are harder to instrument and the wins are harder to demonstrate in a screenshot.

The empirical record makes the same point from the opposite direction. GitHub's Octoverse 2025 round-up reported that developers using Copilot on a bounded JavaScript HTTP-server task completed it 55 per cent faster than the control group (GitHub 2025). METR's randomised controlled trial of experienced open-source developers working in their own mature repositories, by contrast, found a 19 per cent *increase* in completion time when AI assistants were used, even though those developers perceived themselves as 20 per cent faster (Becker et al. 2025). The contrast is the point. Bounded, well-specified work is where Task Delegation pays. Deeply contextual, complex work in a long-known codebase is not, and the developers doing it are unreliable judges of whether the help is helping.

24.3 Code generation is real Task Delegation, with real discipline

Code generation is where Task Delegation comes into its own, and the engineering profession has, on balance, learned to use it well.

A well-specified implementation task has clear inputs, clear outputs and a reviewable artefact at the end. Write a function that parses this format. Add a database migration that backfills this column against this constraint. Implement this endpoint against this contract. These are tasks, not problems. The human assigns the work, the model produces a candidate solution, and the human reviews it with professional rigour, checking correctness, fit with surrounding code, performance characteristics, test coverage and security posture. The model is not a peer in this exchange. It is a fast, capable executor whose output must pass the same bar as any other contribution to the codebase.

The discipline that matters here is specification.

Vague tasks produce vague output. Accepting vague output silently corrodes a codebase. Engineering teams that work well in this mode invest in clear interfaces, strong typing and comprehensive tests, because these are the scaffolding that makes delegated work reviewable. They also invest in their own reading skills. A senior engineer who can no longer read code carefully is not equipped to delegate its production. The reading is the review, and the review is what keeps the system honest.

Code review of human-written code shifts subtly when AI is part of the workflow. Reviewers can use the model to identify likely defects, surface stylistic inconsistencies and check for known vulnerability patterns before they spend their own attention on the change. The reviewer remains the decision-maker. The surface area they can attend to grows. This is still Task Delegation, with the review itself as the delegated task, and it works for the same reason the original generation works: a clear input, a clear output and a human applying professional judgement to the result.

24.4 Autonomous Agency works where the scope holds

Autonomous Agency is the most recent mode to enter the engineering toolkit, and the one most prone to misuse. The human defines an objective and a set of guardrails, and the agent determines its own implementation path. The human reviews the outcome rather than the steps.

The bug-fixing agent is the canonical case, and worth working through carefully because it shows what holds the mode together when it works.

> *Given a failing test, a defect report and access to the relevant repository, a bug-fixing agent investigates the failure, hypothesises causes, writes candidate fixes, runs the test suite and submits a pull request. The engineer does not pair on the investigation. They review the resulting change as they would any other contribution, with the additional knowledge that the agent operated within a defined scope and that the change has already passed the project's automated checks.*
>
> *If the fix is right, it merges. If it misses something a test did not catch, the review catches it. If the agent gets stuck, it escalates with what it tried and where it stopped.*

Three features of this scenario explain why Autonomous Agency is the right operating model rather than Paired Cognition. The objective is unambiguous:

make the failing test pass without breaking the others. The scope is contained: a defined repository, a defined defect, a defined set of automated checks. The cost of a wrong outcome is recoverable: a bad pull request is rejected at review or reverted after merge, and the cost of either is small. Strip any of these features away and the case for autonomy weakens. An ambiguous bug report against a system the agent has never seen, with no test to anchor the fix, is a Paired Cognition problem dressed in agent clothing.

The same logic generalises. Autonomous Agency is appropriate when the objective is clear, the scope is bounded and the cost of a wrong outcome is recoverable. Routine dependency upgrades, well-understood refactors, test generation for legacy code and triage of low-priority issues all fit. It is inappropriate when the objective is ambiguous, when the agent would need to make architectural choices to proceed or when the failure modes are not contained by the test suite and review process.

The guardrails are as important as the objective. Repository scope, branch protection, the inability to merge without human review, rate limits on resource use and clear escalation paths when the agent gets stuck are what keep autonomous work safe. Engineering organisations that adopt this mode without investing in those guardrails accumulate risk faster than they accumulate productivity, and the risk compounds quietly until something visible breaks.

24.5 Operations is the natural home of Embedded Intelligence

Operations is where AI becomes a system component rather than a collaborator. Monitoring, automated remediation, incident triage, log analysis and alerting are continuous activities whose value compounds when AI is wired into the substrate of how the platform runs.

In this mode the AI is not invoked by a human for a specific request. It runs constantly, processing telemetry, classifying anomalies, correlating signals across services and surfacing the small fraction of events that warrant human attention. The site reliability function has been moving in this direction for a decade. Modern models accelerate the trajectory without changing its shape.

The applications are well understood in cloud-native organisations. An anomaly detector that learns the normal traffic profile of each service and flags deviation. A log analysis pipeline that clusters errors and surfaces novel patterns rather than re-paging a team about a known one. An incident triage system that drafts initial assessments by correlating fresh alerts with recent deployments and related historical incidents. A remediation workflow that handles known failure modes, restarting a pod, rotating a credential, draining a node, without paging anyone.

What distinguishes Embedded Intelligence is that the human is not in the loop for routine activity. They set policy, review aggregated outcomes, and intervene when the system surfaces something it cannot handle. The cognitive partnership is at the level of the system, not the individual interaction. This requires investment in observability, in feedback loops that let the embedded intelligence improve over time and in clear ownership of which decisions it is permitted to make and which it must escalate.

The temptation in operations is to over-trust. A remediation workflow that handles ninety per cent of cases cleanly will be allowed, by drift, to handle the ten per cent it should not. The discipline is at design time. The runtime cost of a missing guardrail is measured in incidents.

24.6 Where engineering leads, and where it under-uses what it has

Technology teams are well-positioned for AI adoption in ways other functions are not. They are comfortable with tooling, fluent in iteration and culturally aligned with automation. They adopt new instruments early, share what they learn, and build feedback loops into their workflows by reflex. These are real advantages, and they explain why engineering has produced more visible AI integration than most other parts of the enterprise.

The under-use is more interesting than the lead, and it has a specific shape worth naming.

Engineering teams systematically default to Task Delegation because code generation is the most visible application and the one with the strongest tooling. The models are tuned for it. The IDE integrations are mature. The metrics, accepted suggestions, lines committed, pull requests merged, are easy to count. The function gravitates toward the mode it can measure, and the modes it cannot measure get less attention than they deserve.

Three areas of under-use stand out, and each maps to a class of work that is worth more than any volume of generated code.

The first is high-level design. Architects ask the model discrete questions instead of working through trade-offs in extended dialogue. The work that should be Paired Cognition collapses back into search. Decision records are written without the model having ever held the constraints together. The trade-offs that get captured are the ones the architect happened to think of unaided, which is exactly the set the model could most usefully have pushed against.

The second is cross-system reasoning. Most consequential engineering decisions span repositories, design documents, vendor contracts, runbooks and conversations that no single tool surfaces cleanly. Synthesising across this surface is exactly the kind of breadth a model can supply, and yet it is rarely attempted, because building the context up from scratch each time feels like more work than carrying the synthesis in the head of whoever has been around longest. So the synthesis stays in that person's head, the model never sees the full picture, and the organisation pays the cost of the missing view in the form of decisions that look reasonable inside one repository and incoherent at the seam between two.

The third is the work that engineering tends to treat as outside its function: stakeholder communication, requirements exploration, product discovery, the writing of decision rationale that non-engineers will read. This work is exactly where Paired Cognition produces the most leverage, because it is where the engineering function's reasoning meets the rest of the organisation's. An engineering leader who works through a stakeholder briefing in dialogue with a model arrives at the meeting with the trade-offs framed in the language the meeting requires. An engineering leader who treats stakeholder communication as overhead arrives with a slide deck that the rest of the room has to translate.

The risk of defaulting to Task Delegation is not that the delegated work is bad. The delegated work is generally good and getting better, and the risk lies elsewhere: the higher-value applications never get attempted at all, because the function has built its instincts and its metrics around the mode it can already see. An engineering organisation that uses AI only to generate code is improving its execution speed on whatever it has decided to build. It is not improving the quality of the deciding. DORA's 2025 *State of AI-Assisted Software Development Report* found that AI adoption correlates positively with throughput and product performance but negatively with delivery stability (DORA 2025). AI amplifies whatever discipline, or lack of it, already exists in the delivery system, and a function that has only learned to write code faster will find that the disciplines it has not yet built are the ones AI exposes first.

Over a multi-year horizon, the second effect dominates.

The teams that get the most from AI distribute their use across all four modes deliberately. They reach for Paired Cognition when they are reasoning about systems, products and design. They reach for Task Delegation when they are producing code under specification. They reach for Autonomous Agency when they are clearing well-bounded backlog inside guardrails that hold. They reach for Embedded Intelligence when they are operating production. Without changing what good engineering looks like, the framework changes where the leverage sits, and it makes visible the work the function has been quietly under-investing in for years.

CHAPTER 25

AI in Education: Teaching, Curriculum and Administration

> *Any subject can be taught effectively in some intellectually honest form to any child at any stage of development.*
>
> — Jerome Bruner

Technology teams build the tools and shape the infrastructure that other sectors depend on; education is where the people who will inherit those systems are prepared to use them thoughtfully.

The same teacher who uses a language model to draft differentiated worksheets on Monday morning will, by Friday afternoon, sit down to mark a stack of essays in which the same technology has been used to undermine the assessment she designed. She is not living a contradiction. She is doing her job inside a sector that contains, in the body of a single educator, more layers of cognitive work than almost any other.

Few sectors have welcomed AI with such evident enthusiasm in some quarters and met it with such pointed scepticism in others. The reactions are usually framed as a debate about technology. They are more accurately a debate about which work, in this sector, is even the same kind of work.

Teaching, learning, curriculum design and the administration of schools and universities are not a single activity. They are a layered set of practices, each with different demands, different ethical commitments and different relationships to expertise. The four operating models, distinguished by where decision-making responsibility sits, offer a way to think about these layers without collapsing them into a generic question of whether AI is good or bad for education.

25.1 The sector contains every operating model at once

Most industries lean toward one or two operating models. Education uses all four, often in the same week, and frequently in the hands of the same educator.

A head of mathematics planning a unit on quadratics on Tuesday is in Paired Cognition with an AI that helps her think through the sequence and the differentiation. On Wednesday she might delegate the production of practice question sets against a clear specification, in Task Delegation. Her school's timetabling engine, optimising room allocation overnight, is Embedded Intelligence. The plagiarism screen running quietly across submitted assignments is Embedded Intelligence too. None of these are the same kind of partnership, and treating them as if they were is what makes the conversation about AI in schools so consistently muddled.

The decisive axis, as in every other industry, is decision-making responsibility. Where does the judgement actually sit, and what is the AI doing relative to that judgement? In education, the answer changes not just from task to task but from one part of an educator's day to another.

What makes the sector distinctive is that the central act, teaching itself, sits inside a relationship. A lesson is not simply the delivery of content. It is a sustained interaction between a teacher who knows a particular cohort and the students within that cohort. Anything that touches that central relationship has to be understood differently from the surrounding work that supports it.

The framework is useful here precisely because it locates where AI strengthens that relationship and where it must not intrude. It does not promise that the central act of teaching can be partially automated. It promises something more modest and more practical: a way to identify which of the surrounding tasks can be reorganised, and on what terms.

25.2 Lesson planning is the clearest case for Paired Cognition

Lesson planning is exhausting in a way the public conversation about teaching rarely captures. A teacher plans across multiple constraints at once: the curriculum standards she is responsible for, the prior knowledge of the class, the specific learners with adjustments or extension needs, the pacing of the term, the resources available and her own pedagogical preferences. Holding all of this in mind while drafting from scratch is the kind of sustained reasoning that Paired Cognition was designed to support.

In a Paired Cognition mode, the teacher articulates the goals and the constraints, and the AI proposes structures, activities, sequences and worked examples. The teacher then critiques, redirects, rejects, refines. The mental model is shared because the AI has been given enough context to produce material that is at least adjacent to what the class needs. The pedagogical judgement remains entirely with the educator. The cognitive load of generating options drops substantially.

> *A Year 9 English teacher is planning a four-lesson sequence on a contested non-fiction text. She externalises into the AI what she knows about the class: two students with reading-comprehension adjustments, three who finish everything early and disengage, a recent cohort discussion that revealed surprisingly strong views on the text's subject matter. She lays out the standards she is working against and the sequence's place in the term.*
>
> *The AI proposes three different shapes for the sequence, each with rationale. The teacher rejects two and works with the third for an hour, refining the questions, swapping in a primary source she had been waiting to use, redesigning the formative check at the end of lesson two. The output is unmistakably hers. The hour spent producing it would, two years ago, have been three hours and a worse plan.*

The time saving here is not anecdotal. In a randomised trial run by the Education Endowment Foundation and NFER in 2024, secondary science teachers using ChatGPT alongside a brief usage guide spent 56.2 minutes per week on lesson preparation, against 81.5 minutes for the control group: a 31 per cent reduction, with no noticeable difference in lesson quality assessed by an independent panel (Education Endowment Foundation and NFER 2024). What the trial measured was Paired Cognition working as intended: the teacher's judgement preserved, the surrounding labour reduced, the artefact at least as good.

Differentiation design works the same way. A teacher who needs three reading versions of the same source text, or four levels of mathematical scaffolding for a single concept, can use the AI as a thinking partner rather than as a producer of finished artefacts. Assessment design also benefits, particularly the slow work of writing rubrics that genuinely capture the intended learning rather than only what is easy to mark.

What this collaboration does not do is replace the educator's knowledge of why particular activities work for particular students. The AI can suggest that a Year 9 cohort might enjoy a debate format. It cannot know that this specific class includes a student who will not speak in a debate and another for whom debating is the only time he engages. That knowledge belongs to the teacher and shapes every adjustment she makes to the AI's suggestions.

25.3 Curriculum design needs breadth from AI and depth from people

Curriculum design operates at a longer timescale than lesson planning and pulls in more sources. A curriculum leader designing a new senior subject offering, or revising a primary school's literacy approach, draws on academic research, official curriculum frameworks, professional standards, the school's strategic priorities and the realities of the staff who will teach the work. AI systems can synthesise large bodies of public material in ways that a single human cannot easily replicate.

Used well, this allows curriculum leaders to scan literature, compare approaches across systems and surface alternative sequences far more rapidly than before. The AI brings breadth. The educator brings depth: the pedagogical judgement to evaluate which approach actually fits the learners in this school, the institutional context that shapes what is possible and the deep knowledge of staff capability that determines what will be sustained in practice.

This is again Paired Cognition rather than delegation. Curriculum is not a document but a coordinated set of decisions about what is worth knowing, what counts as evidence of learning and how knowledge is built across years. AI can accelerate the gathering of options and the structuring of arguments. The decisions themselves rest with people who carry professional responsibility for them and who will live with the consequences in the classroom.

The Reflection phase matters here in a way it does not for a single lesson. Curriculum decisions need to be captured as institutional memory, not just as documents. Why was this sequence chosen over the alternative considered? Which assumptions about staff capacity were load-bearing? When the next leader inherits the work, the knowledge base should let her see the reasoning, not just the artefact.

25.4 Some assessment work fits Task Delegation; some does not

Some assessment work is genuinely well suited to Task Delegation. A teacher with a clear marking rubric and a stack of structured short-answer responses can ask an AI to draft preliminary feedback against the rubric, which she then reviews, adjusts and signs off. The criteria are explicit, the task is bounded and the educator's review is not optional but constitutive of the work.

The same applies to producing model answers, generating practice questions aligned to particular standards, drafting parent-facing comments from teacher notes and turning observation evidence into formal report language. These are tasks where the educator has decided what should be said and the AI assists in saying it more efficiently. The human remains the author of the judgement.

The most striking evidence for this kind of supported delegation comes from outside the classroom proper. A 2024 randomised trial of Stanford's Tutor CoPilot (running with more than 900 tutors and 1,800 K-12 students from underserved communities) found that students whose tutors used the AI assistant were 4 percentage points more likely to master a maths topic, with the largest gain (9 points) accruing to students working with lower-rated tutors (Wang et al. 2024). The AI lifted the floor more than it raised the ceiling. That asymmetry is itself the operating-model claim. Task Delegation, properly designed, does not replace the human in the relationship; it supplies the less experienced practitioner with scaffolding the experienced practitioner already carries internally, and the gain accrues exactly where the practitioner had least to draw on.

What this model requires, and what is easily lost in busy schools, is the discipline of review. Feedback that is delegated to AI and then forwarded without genuine engagement is not Task Delegation in any meaningful sense. It is abandonment of the assessment relationship. The framework only works if the review step is real, and if the educator is willing to override the AI when its draft misses what actually matters about a particular student's response.

There is a quieter problem worth naming here. Assessment carries two purposes at once, and the second one, which is feedback to the teacher about her own practice rather than to the student about hers, is the one that gets quietly lost when marking is automated. A teacher who marks a class's work herself notices, in the marking, things she would not have noticed any other way: that two thirds of the cohort misunderstood the question stem, that a particular misconception is more durable than she had thought, that her best student has stopped trying. AI-assisted marking does not have to lose this signal, but it loses it by default. Recovering it requires deliberate review, not glance-and-sign-off.

> *A Year 11 chemistry teacher uses an AI to draft feedback on a stack of structured-response papers against a five-criterion rubric. The drafts are competent. She reviews them while sitting with the actual scripts, not just the AI's summaries. On three of the papers she rewrites the feedback substantially. On one she changes the band judgement, because the script does something subtle the rubric does not quite capture and the AI has rewarded surface fluency over conceptual accuracy.*
>
> *She notices, in the act of reviewing, that the cohort has a shared confusion about a concept she had assumed was settled. She rewrites the next lesson. None of this would have happened if the feedback had been forwarded without review.*

25.5 Administration is the natural home of Embedded Intelligence

Schools and universities run on administration. Timetabling, room allocation, attendance tracking, enrolment management, reporting cycles, compliance returns, leave management and the constant flow of communication with families all consume substantial staff time. Much of this work is rule-bound and repetitive. It is the natural home of Embedded Intelligence.

In this mode, AI is a component inside a larger system rather than an interlocutor. A timetabling engine that uses AI to optimise across constraints, a reporting platform that drafts standard sections from structured data, a records system that flags anomalies for review, all sit comfortably here. Although the educator does not collaborate with these systems in any active sense, they simply produce better outputs than the previous generation of tools did.

Embedded Intelligence is where many institutions will find the largest and least controversial gains. It frees educator time for the work only educators can do.

The risks are mostly about governance. Ensuring that the data flowing through these systems is handled appropriately, that the rules encoded into them reflect institutional values and that the people whose lives are shaped by their outputs have meaningful recourse when the system gets something wrong. A timetabling engine that quietly disadvantages students who take less common subject combinations is not a neutral piece of infrastructure. Someone has to be accountable for the rules it is running, and that accountability needs a human address inside the institution.

25.6 What stays with the educator is not a list but an argument

It is tempting, when defending the educator's role, to produce a list of human qualities a system cannot replicate. Care. Empathy. Judgement. The list reads as nostalgic rather than analytic, and it loses ground every time the technology improves on a quality the list claimed for the human. The argument has to be made differently.

Pedagogy is the first part of the argument, and pedagogy here means something narrower than general knowledge of how learning works. It is the disciplined art of helping a particular person learn a particular thing in a particular moment, drawing on what that person already knows, what is currently confusing them and what will move them forward without overwhelming them. A system that has not watched the learner cannot do this. A system that has only the learner's outputs cannot do it either, because the outputs are downstream of the misconception, not the misconception itself. Pedagogy lives in the diagnostic act that precedes the explanation, and the diagnostic act depends on knowing the learner well enough to read the gap between what she has produced and what she is actually thinking.

The learning relationship is the second part. Learning, for the kind of work schools and universities do, is not the transfer of information. It is the slow business of taking on a way of thinking, often against the learner's own resistance, and the energy that carries the learner through that resistance is borrowed from someone she trusts. A teacher who has held expectation steadily through a difficult term, who has shown a student that she believes the student capable of the work the student doubts she can do, is doing something a system cannot simulate. The simulation fails not because the words are wrong but because the trust is what was load-bearing, and trust is a relational fact, not a textual one.

Knowing individual students is the third part, and probably the deepest. A teacher who has watched a child for a term knows when withdrawal signals difficulty at home, when overconfidence is masking confusion and when a quiet student is finally ready to take a risk. None of this is encoded anywhere an AI can access. It is read from posture, from the timing of a question, from a comment made on the way out of the room. The teacher carries this knowledge as the ground on which every adjustment to a lesson, every redirection of a struggling student, every decision about who to push and who to leave alone today, actually rests. Without it, every other intervention is poorly targeted, however polished.

These three are not independent. Pedagogy without the learning relationship becomes technique applied to strangers. The relationship without pedagogy becomes warmth without progress. Both depend on knowing the learner. Together they sit close to the irreducible centre of the work, and the framework is useful precisely because it leaves that centre intact. It identifies the surrounding work that AI can support and clarifies the conditions under which that support is appropriate. It does not claim that the central act of teaching is one of those tasks.

There is a tendency, when a new technology arrives in a sector, to ask which of the human roles it can replace. In education the more useful question runs the other way. Which parts of the educator's week are not actually about the learning relationship at all, and could be reorganised so that more of her attention is available for the parts that are? The answer turns out to be substantial, and almost none of it touches the relationship itself.

25.7 Academic integrity is an authorship question, but the authorship is not always clean

Academic integrity is the area where the framework offers the sharpest clarification, and also the area where clarification has to stop short of resolution. The question is not whether AI is permitted in education. It is whose work the AI is producing, and what the work is supposed to be evidence of.

When AI assists an educator with lesson planning, rubric drafting or report writing, it is producing educator work. The educator's professional responsibility is to use her tools well, and AI is a tool. There is no integrity question because the educator is not being assessed. Her output is judged by whether it serves her students, not by whether she produced it unaided.

When AI produces student work, the situation is fundamentally different, because the purpose of student work is usually to develop and demonstrate the student's own thinking. AI that substitutes for that thinking undermines the purpose of the assessment, regardless of how polished the output looks. Whether and how students may use AI is then a matter of pedagogical design, and the design is the integrity question.

The clean version of the distinction is too clean. A great deal of student work sits in between. A student who uses AI to find a counterargument she had not considered, then engages with it in her own voice, has used the tool in a way that develops her thinking rather than substituting for it. A student who uses AI to translate her own argument from rough notes into polished prose has done something more ambiguous: the thinking is hers, the sentences are not. A student who has the AI generate the argument and then edits the surface of it has produced something that looks like her work and is not. These are different cases, and a policy that cannot tell them apart will either over-permit or over-prohibit, depending on which way the institution leans.

The honest position is that the line between developmental use and substitutive use is real but moves with the task. For an early-secondary student learning to construct a paragraph, having AI compose the paragraph defeats the entire point. For a postgraduate refining a literature review, having AI suggest related work she might have missed is no different from a librarian doing the same thing twenty years ago. The same activity is appropriate at one stage and corrosive at another, and that is a pedagogical judgement, not a technology judgement.

What this means in practice is that integrity policy cannot be a single rule. It has to be embedded in the design of each assessment. The teacher who sets the task is the only person in a position to say what the task is meant to demonstrate, and therefore the only person who can say which uses of AI undermine that demonstration and which do not. Institutional policy provides the frame; the unit and the task fill it in.

Educators can use AI for their own work without that fact alone telling students anything about what is permissible in theirs. The two domains have different purposes and call for different rules. A teacher who uses an AI to draft a lesson plan is not modelling permissible behaviour for a student writing an essay, any more than a chef using a thermometer is modelling permissible behaviour for a student in a sensory cookery exam. The asymmetry is not hypocrisy. It is a feature of the relationship: the educator's tools are judged by their effect on the student's learning, and the student's tools are judged by their effect on her own.

The framework does not resolve every dilemma, though it does sharpen the question. The integrity issue, properly stated, is less about whether the technology exists than about what the assessment is for, and whether the form of the task still measures what it claims to measure now that the technology exists. That is a pedagogical question, and educators are equipped to answer it, one task at a time.

25.8 Equity is a complication that runs in several directions at once

Equity considerations are usually mentioned in education-and-AI conversations as a closing caveat. They deserve more than that, because the equity picture is genuinely complicated and the complications run in different directions.

The first direction is the obvious one. Differential access to AI tools across student populations matters because it changes what learners can produce and what supports they can draw on outside school. A student with a paid frontier model at home and a parent who knows how to use it has resources another student does not. An assessment regime built on the assumption of no AI access, in a world where some students have it and others do not, is no longer measuring what it thinks it is measuring. The integrity of the assessment and the equity of the assessment are, in this respect, the same problem viewed from two angles.

The second direction cuts the other way, and is less often acknowledged. AI may, for some students, prove a more significant equaliser than much of what education has produced in a generation. A student with a learning difficulty who has never been able to externalise her thinking onto a page can now do so. A student whose first language is not the language of instruction can write at the level of her thinking rather than at the level of her vocabulary. A student whose family cannot afford tutoring can have a patient explainer available at every hour. To frame AI access purely as a privilege the advantaged enjoy is to miss that, used carefully, it removes barriers that previously kept particular students from showing what they knew.

These two directions do not cancel out so much as make the design problem harder. An institution that bans AI in pursuit of a level playing field may be removing a tool that was, for some students, the first thing that let them play the field at all. An institution that permits AI without thought may be widening the very gap it imagines itself closing, because the students who already had every advantage are also the students best placed to use the new tool well.

The point is not that every student needs identical access to identical tools. It is that the policy decision sits inside a real distributional problem, and the distributional problem has to be reasoned about explicitly rather than waved at. Provision of school-managed AI access, training in how to use the tools to support rather than substitute for learning and assessment design that accounts for the realistic distribution of access in the cohort are all parts of the response. None of them on its own is sufficient. The framework does not resolve this; it does insist that the question is asked workflow by workflow rather than answered by a single institutional rule.

The operating model considerations also differ between higher education and K-12, and these differences interact with the equity picture in ways worth naming. In higher education, students are adults expected to manage their own learning, and the question of permitted AI use can be devolved further to the discipline and the unit. The norms of historical scholarship are not the norms of software engineering, and a sensible institution lets each discipline articulate its own.

In K-12, particularly with younger learners, the school carries a stronger duty of care over how children encounter these tools and what habits of thinking they form while doing so. A ten-year-old who learns to outsource the act of writing before she has learned to write is not the same kind of educational case as a postgraduate using a language model to refine a literature review. The framework applies in both settings. The settings themselves are different, and the equity considerations that bind in each are different too.

25.9 What institutions gain is a coherent position, not a final answer

What institutions gain from working with the framework is clarity. They can articulate, to staff and to families, where AI is being used in their operations, where it is being used by educators to support teaching, where students may use it in their learning and where it must not substitute for human teaching at all.

That articulation does not solve every dilemma the technology brings, and new ones will arrive faster than policy can be written about them. The benefit of a coherent position is not that it answers every future question but that it provides the institution with a stable place to stand while the questions are being asked.

A school that can name, for any given activity, which operating model is in play and why, is in a different conversation from one that cannot. It can talk to parents about what teachers are doing with AI without conceding ground on what students are not. It can integrate Embedded Intelligence into administration without that integration becoming a debate about teaching. It can let educators experiment with Paired Cognition in their planning without that experiment migrating uninvited into the assessment relationship.

Education is the last of the seven industries this part has worked through, and the picture it leaves is the picture the whole part has been assembling. Each industry runs all four operating models simultaneously. Each one has work that responds best to sustained partnership and work that is better delegated, automated or embedded. Each one has at its centre something that the framework leaves alone: the underwriter's judgement about a non-standard risk, the clinician's responsibility for a patient, the lawyer's professional obligation, the educator's relationship with a learner. The framework's contribution is not to dissolve those centres but to organise the work around them, so that the central judgement is made with more support and less surrounding noise.

The harder question, once an organisation can name the operating model in play for any given workflow, is what happens when those workflows meet. The educator whose lesson planning runs on Paired Cognition is not insulated from the timetabling engine that runs on Embedded Intelligence, the assessment platform that drafts feedback under Task Delegation or the analytics service that quietly classifies which students are at risk. Each of these touches the others, and the points where they touch are where coherence is built or lost. That is a problem that no single industry chapter can resolve, because it is not an industry problem. It is the organisational problem that Part V begins.

The technology will keep moving. The relationships at the centre of education, and at the centre of every other field this part has examined, will not. The institutions that come through the next decade well are the ones that can tell the difference, in detail and at scale, between the work that AI can usefully reorganise and the work that has always been, and remains, irreducibly human. Telling the difference inside a single workflow is the discipline Part IV has developed. Telling it across an entire organisation is the work that begins now.

Part V

The Organisation and Its People

The enterprise-level questions: orchestration across operating models, governance, why AI adoption fails, the human side of partnership and the trajectory ahead.

Chapter 26

Orchestration: Composing AI Methods Across an Organisation

The purpose of a system is what it does.

— Stafford Beer

An organisation that has chosen the right operating model for each of its workflows still has work to do. The question is no longer which model is best for any single workflow. It becomes which model belongs where, and how the chosen models behave when they meet at the seams between teams, systems and the rest of the business.

The four models describe relationships within a single workflow. Orchestration describes the composition of those relationships across an entire enterprise. It is the discipline that determines whether the others compound or collide.

Part V takes the framework to the organisational level. Orchestration first, then governance, adoption, the human side of partnership and the trajectory of cognitive work itself. Each of the chapters that follow assumes the same starting point: an organisation running more than one operating model, in more than one part of the business, at the same time. That starting point is not automatic. Most organisations have to be talked into it, and the talking begins here.

26.1 Why is monoculture so seductive?

The most common failure pattern in organisational AI deployment is not picking the wrong model for a workflow. It is picking one model and applying it everywhere. Monocultures simplify procurement, training and governance. They also fail in characteristic ways, and those ways are visible only after significant investment has been made.

Consider two firms at opposite ends of the spectrum.

A mid-sized professional services firm decides its AI strategy will be built around Autonomous Agency. Document review, drafting, due diligence summaries and routine client correspondence are all routed through agentic systems that operate with minimal human involvement. The first six months produce impressive throughput numbers, and the leadership team writes the case study.

The second six months produce a quiet accumulation of subtle errors that nobody caught, because nobody was in the loop to catch them. By the time a regulator surfaces the pattern, the firm has built a defensive workstream large enough to erase the original gain. The leadership team writes a different case study, internally.

The opposite monoculture fails differently, and more expensively.

A regional bank makes Paired Cognition the universal standard. Analysts and officers spend their days in deep partnership sessions across every category of work, including the work that does not need it. Status updates that used to take twenty minutes now consume two hours of partnership time. Routine compliance summaries, which a Task Delegation pattern would draft adequately in seconds, instead occupy senior cognitive effort.

The quality of output is unusually high across the board. The throughput is catastrophic. The bank is paying its best people to think carefully about work that did not require careful thinking.

Both organisations are failing for the same underlying reason. They have treated AI methodology as a single choice rather than a portfolio of choices.

Although the four operating models can look like competing answers to the same question, they are answers to different questions altogether. A healthy organisation will be running all four simultaneously across different workflows, and the maturity of its AI practice can be read directly from how cleanly the four are kept distinct, deployed where they fit and connected at the points where they touch. McKinsey's March 2025 *State of AI* (McKinsey & Company 2025c) found that AI high performers, the small share of organisations consistently capturing EBIT impact, were nearly three times as likely as their peers to have fundamentally redesigned their workflows for AI, and that workflow redesign had the single largest correlation with value capture among the twenty-five organisational practices the study tested. The portfolio view is what makes that redesign tractable. An organisation that has not separated its workflows by cognitive character cannot redesign them differentially, and undifferentiated redesign is the form monoculture takes when it dresses itself up as transformation.

That last clause is where orchestration begins.

26.2 Governance lives at the seams

Once the monoculture trap is understood, the next problem is governance. If different workflows use different models, who decides which model applies where, and how are those decisions kept coherent over time? This is the question that turns orchestration from a description into a discipline.

The decision rights question is genuinely hard. Operating model selection cannot live entirely with the technology function, because the choice depends on the nature of the work and the risk appetite of the business. It cannot live entirely with the business function either, because the choice depends on capabilities of the underlying systems and on what those systems can be configured to do safely. It is an inherently shared decision, and shared decisions need explicit structure.

Most organisations that orchestrate well end up with a small central capability that holds the framework, sets the criteria for model selection and reviews proposed deployments against those criteria. The actual decisions are made in the business. They are made within a structure that prevents drift, contradiction and duplication.

The harder problem is what happens at handoff points. A loan application processed by Embedded Intelligence eventually reaches a credit officer working in Paired Cognition. A research brief produced through Paired Cognition is handed to an Autonomous Agency pipeline that drafts the deliverable. A monitoring alert raised by Embedded Intelligence triggers an investigation that proceeds as Paired Cognition until a discrete sub-task is sub-delegated to Task Delegation.

Each transition is a potential point of failure. Each one requires explicit design.

This is the orchestration challenge in its purest form. Inside any one model, the discipline is about doing that model well. Between models, the discipline is about ensuring that what one model treats as output another can treat as input, without losing information, mistaking confidence for certainty or letting accountability fall into the gap.

26.3 What is the cognitive shape of the work?

Orchestration begins with an inventory. Not a strategy document but a map of the actual workflows that constitute the business, the systems they touch and the cognitive character of the work being done at each step. Without this map, model selection becomes opinion rather than analysis.

A useful inventory begins with the nature of the cognitive work itself, because everything else follows from it. Is the work judgement-heavy, repetitive, exploratory, transactional? That answer constrains the model selection more than any other variable. It then needs to be qualified by what an error would cost and whether the cost is recoverable, because the same kind of work attracts different operating models depending on how much room it has to be wrong. Volume sits underneath both. A judgement-heavy task that occurs twice a year is a different orchestration problem from one that occurs twice a day, even when the cognitive character is identical, because the economics of partnership shift with frequency.

Worked through the inventory, these factors yield a defensible matching to operating models. Judgement-heavy, low-volume, high-consequence work tends toward Paired Cognition. Repetitive, high-volume, bounded-consequence work tends toward Task Delegation or Embedded Intelligence. Exploratory work with well-defined outputs tends toward Autonomous Agency. The categories are not rigid, but the analytic discipline matters more than any single placement.

The inventory exercise itself often surfaces workflows that nobody had previously named as workflows. Internal handovers between teams. Informal review cycles. The long tail of administrative work that lives in email threads and shared drives. These are precisely the places where Embedded Intelligence quietly delivers most of its value, and where the absence of orchestration shows up as friction nobody can locate.

> *A regional insurer building its first orchestration map began with the obvious flagships: underwriting on the Paired Cognition side, claims triage on the Autonomous Agency side. The exercise took on a different character once the team started tracing what happened between the named workflows.*

The submission queue handed to underwriters was being assembled by three overlapping spreadsheets and a daily email. The endorsement letters that returned to brokers were being drafted in a tool that one team had bought and another team did not know existed. Two of the most consequential AI choices the insurer needed to make sat in workflows that had never been named.

The map is not the strategy itself but the surface on which any coherent strategy has to be drawn, and the absence of that surface is why so many AI strategies float free of the workflows they are meant to address.

26.4 Composition lives inside workflows as well as between them

The four operating models are not just deployed alongside each other. They compose. A single workflow can incorporate multiple models in a structured way, and understanding these compositions is central to orchestration as a practice.

Paired Cognition is the most common composer, and its dominant composition pattern is worth describing carefully because it is widely misunderstood.

During the Execution phase of a Paired Cognition workflow, the human and the primary AI agent routinely identify discrete units of work that do not require ongoing dialogue to perform. A specific draft section. A bounded analysis. A transformation of one form of content into another. These units can be sub-delegated to Task Delegation, where sub-agents handle specific tasks against a brief while the primary partnership continues at the orchestrating layer. The structure was named earlier as sub-delegation. At the organisational scale it becomes a composition pattern in its own right.

This nested composition is powerful because it preserves the strengths of each model. The primary agent maintains the dialogue, the judgement and the synthesis. The sub-agents do bounded work efficiently. The human reviewer engages with the partnership conversation rather than with every sub-task, and the overall throughput is far higher than pure Paired Cognition would deliver.

The shared mental model carries through. The sub-task is briefed by an agent that already understands the context, intent and standards of the surrounding work. The cold-brief problem that limits Task Delegation in isolation is largely sidestepped, because the brief is being written by something that already holds the context the brief would otherwise have to articulate. This is why nested composition does not collapse into ordinary Task Delegation. The container is a partnership; the inner work is delegated; the result inherits the partnership's grasp of what the work is for.

Boundary compositions matter as much as nested ones. Autonomous Agency systems frequently feed their outputs into partnership workflows for high-stakes review. Embedded Intelligence often surfaces signals that trigger partnership engagement. These are the seams of the organisation, and how cleanly they are designed determines whether the overall system feels coherent or jagged to the people working inside it.

The diagnostic to apply at every seam is simple. The receiving model needs to be able to treat what arrives as a clean input. If a Paired Cognition workflow inherits an Autonomous Agency output, the partnership has to know what was done, what was assumed and what remains uncertain. If an Autonomous Agency pipeline inherits a Paired Cognition decision, the parameters of the decision have to be expressed in terms the pipeline can act on without further interpretation.

A seam that fails this test is not a seam. It is a hidden translation step that some human will have to absorb in order for the work to keep moving.

26.5 Governance shifts from monitoring behaviour to monitoring composition

Governance designed for a single operating model rarely survives contact with a multi-model environment. Oversight structures need to span the portfolio, and they need to do so without collapsing into either rubber-stamp committees or bottlenecks that block delivery.

The most useful pattern is to govern at the seams rather than within each model. Govern the joints, not the limbs. The internal mechanics of a well-designed Paired Cognition workflow do not need committee oversight. What needs oversight is the decision to deploy Paired Cognition rather than Autonomous Agency for a particular use case, and the design of the handoff points where one model gives way to another.

This is the move that distinguishes mature orchestration from immature versions. Immature governance tries to inspect every output. Mature governance inspects the structure of the system and trusts the operating models to behave as designed within their assigned domains. The shift is from behaviour to composition. It changes what a governance committee is for.

> *A healthcare network running multiple models illustrates the principle in operating form. Clinical decision support runs as Paired Cognition, with the clinician retaining sole decision-making responsibility. Image triage runs as Autonomous Agency with human review on exceptions. Scheduling, intake and administrative communication run on Embedded Intelligence inside the operational systems.*
>
> *The governance committee does not review individual diagnoses or individual rosters. It reviews the criteria for moving a workflow between categories, the design of the handoff where image triage results enter the clinician's reading queue and the escalation paths when a workflow's assumptions stop holding.*

The committee's product is not approval. It is coherence.

The next chapter develops the discipline of governance in detail, including the three-tier control framework and how the Basel three lines of defence transposes onto AI work. The point in this chapter is structural rather than mechanical. Orchestration is what makes governance tractable. Without orchestration, governance becomes a list of every AI system in the organisation. With orchestration, governance becomes a structured view of how those systems compose.

26.6 The recurring orchestration mistakes are recognisable in advance

Monoculture is the most visible failure of orchestration, but it is rarely the only one running in an organisation that has not built the discipline. The other failures are harder to see because they hide behind local successes. They are worth naming clearly so that they can be caught before they consume a programme.

Most of them begin where monoculture leaves off. An organisation that has avoided picking a single model can still fail by failing to coordinate. Individual teams adopt operating models that suit their local context, no central function tracks the resulting portfolio, and the organisation ends up with five different vendors solving overlapping problems, three different definitions of what "AI assistance" means and no shared language for discussing AI work across the executive team. The monoculture problem has been replaced by an uncoordinated polyculture, which fails for almost opposite reasons but produces a similar outcome: the portfolio does not compose.

Inside that uncoordinated portfolio, governance gaps at handoff points become the next failure. Each model is governed adequately within its boundary. Nobody owns the transitions. Errors propagate across these gaps and accumulate in the spaces where no single team has accountability, and they surface as outage post-mortems, regulatory enquiries or quiet customer attrition. The connection to coordination failure is direct. A central function that does not track the portfolio cannot see the seams, and a seam that nobody can see is a seam that nobody will design. Deloitte's 2026 TMT Predictions report (Deloitte 2026c) observed that only one in five companies had a mature governance model for autonomous AI agents, and argued that orchestration governance was the variable separating successful deployments from those that stalled: clear roles, defined accountability, fallback routes between agents. The pattern that report describes is the pattern this section names: governance that stops at the boundary of each model is not governance of the system, and a multi-agent system without seam-level accountability accumulates exactly the failures the Deloitte data flags.

Coordination failures and seam failures together encourage a third pattern, which is the default model trap. Organisations that lack a shared framework for choosing pick by inertia instead. Whichever model was first deployed successfully becomes the model applied to every new use case, often because partnership is what the team knows how to deploy or automation is what the vendor sold last quarter. New workflows that should have been Embedded Intelligence become Paired Cognition. The result is over-engineered solutions to simple problems and under-engineered solutions to complex ones, and the cause is the same absence of central framework that produced the coordination failure in the first place.

The final failure is the one that ties the others together over time.

Orchestration is treated as a one-time exercise. The map gets drawn, the criteria get written, the committee gets stood up and then the framework ossifies while the underlying capabilities, organisational priorities and workforce fluencies move on. An orchestration framework that was correct two years ago will be wrong now. Without a discipline of revisiting model assignments, the coordination problem returns, the seams drift, and the default model trap sets in around whichever pattern the framework once endorsed.

The pattern across all four is the same. Each is a place where the absence of orchestration is invisible until something goes wrong, and each compounds the others. Orchestration is one of those disciplines whose presence is felt as the absence of a class of problems. That makes it easy to underinvest in.

26.7 Orchestration is built deliberately or it is not built at all

Orchestration is a skill that organisations build deliberately or fail to build at all. It rarely emerges from existing functions without explicit investment. The technology function alone lacks the business context. The strategy function alone lacks the technical depth. The operations function alone lacks the methodological vocabulary.

The most common home for the capability is a small group sitting between strategy, technology and operations, often reporting to a senior leader who holds responsibility for organisational transformation rather than for any single discipline. The group is small because its leverage comes from framework and coordination rather than from delivery. A team of around four can often orchestrate the AI work of a multinational; a team of forty rarely can, because at that scale orchestration starts to behave like delivery and loses the perspective that gives it value.

The capability matures along a recognisable arc. Early stages focus on inventory and language: simply naming the operating models in use across the organisation and creating a shared vocabulary for discussing them. Middle stages introduce explicit model selection criteria and handoff design, and move governance toward composition rather than behaviour. Late stages embed orchestration thinking into the rhythms of strategy, planning and governance, so that the framework becomes invisible because it is everywhere.

This arc connects directly to the strategy work covered in Part III. Orchestration is the operational expression of AI strategy. A strategy that names ambitions without specifying which operating models will realise them is incomplete. An orchestration framework without strategic direction becomes a sophisticated map of activity that does not add up to direction.

The organisations that get this right do not treat orchestration as a separate workstream layered on top of their AI investments. They treat it as the connective tissue that makes those investments cohere into something larger than the sum of their parts. They staff it accordingly, they protect its independence from any single function, and they expect its value to compound over years rather than quarters.

The chapters that follow take orchestration as given and turn to the questions it makes possible. Governance, in Chapter 27, applies the orchestration view to oversight, accountability and control. The chapters after that examine what fails in adoption, what changes for the people inside teams that work this way, and where the trajectory of cognitive work points next.

None of those questions resolves cleanly inside a single operating model. All of them resolve in the spaces between.

CHAPTER 27

Governance: Oversight, Accountability and Control

> *If men were angels, no government would be necessary. If angels were to govern men, neither external nor internal controls on government would be necessary.*
>
> — James Madison

A senior associate in a regional law firm opens a matter she has not handled before. She has access to several AI tools, a knowledge base her firm has been building for two years, and a workload that does not allow her to think about any of this for long. Before she begins the work, she has to answer a question that nobody has answered for her: whether she is permitted to use AI on this matter at all, which models, with which client material, and under what conditions of review. Every morning, with every new file, she answers it again, alone.

Governance is what removes that question from her morning.

This is the inversion most organisations miss. Although governance is routinely cast as the brake on adoption, what it actually does, when designed well, is make adoption possible without depending on the private courage of individual users. Without it, every act of using AI is an isolated decision under uncertainty, and reasonable people responding to uncertainty default either to caution or to undisclosed use. Both outcomes are bad for the organisation, and both are bad for the practitioner.

The chapter that follows takes governance seriously as a system that lowers the cognitive cost of every individual decision to use AI, rather than as paperwork that exists to satisfy a regulator. The mechanics are conventional in shape. The application is what changes.

27.1 What does good governance actually cover?

Six concerns interlock to form a governance system worth the name. Accountability structure provides a clear answer, for any given AI-supported decision, to the question of who is responsible if it goes wrong. Oversight is the mechanism by which decisions are observed and reviewed at appropriate cadence. Control is the framework that distinguishes what humans must approve, what AI may decide, and what the system runs on its own.

Three further concerns are operational. Audit capability is the ability to reconstruct, after the fact, what the AI did, what data it used, and what reasoning it produced. Escalation paths describe how a user, a reviewer, or a system itself can raise an issue and have it acted upon. Incident response is the standing procedure for handling failures, near-misses, and external complaints with appropriate speed and rigour.

These six concerns are not new. They are the same elements mature organisations already apply to financial controls, model risk and operational risk. The Basel-derived three lines of defence model, in which the business owns risk, a second function challenges and controls it and audit independently assures it, transfers cleanly to AI. What changes is the object of governance, not the discipline.

The discipline therefore exists, and most organisations have a version of it operating elsewhere, so the challenge is not invention but recognition: AI use does not slot neatly into a single existing risk regime, and the governance shape required differs depending on which operating model is in play.

27.2 Governance requirements differ by operating model

The single most important governance insight is that the four operating models carry genuinely different governance requirements. Treating them uniformly is the most common structural mistake organisations make. A control framework calibrated to Autonomous Agency will throttle Paired Cognition into uselessness. A framework calibrated to Paired Cognition will leave Autonomous Agency dangerously exposed. A framework that ignores Embedded Intelligence as not really AI will miss the place where decisions were already made at design time and now run unsupervised.

In Paired Cognition, the human and the AI are jointly engaged in thinking. Responsibility is shared during the work, but accountability for the resulting decision sits unambiguously with the human. Governance therefore concentrates on three things. The human must be competent to partner; operating model literacy is itself a governance condition. The knowledge base they work from must be current and trustworthy. The artefacts they produce must be reviewable in the sense that a colleague, an auditor, or a regulator can later understand what was decided and why.

> *A senior credit officer working a complex limit increase with an AI partner produces a memo that captures her reasoning, the references the AI surfaced, and the points where she set aside the AI's suggestion and chose differently.*
>
> *Two years later, when a regulator reviews the file, the trail is intact not because anyone has been forced to document for documentation's sake, but because the work itself was conducted in a form that is reviewable. The governance was inside the work.*

In Task Delegation, the human briefs the AI, receives an output, reviews it and signs off. Accountability is even cleaner: the reviewer accepts the work and is answerable for it. Governance here focuses on review quality, on ensuring delegation is matched to risk, and on preventing the slow drift in which review becomes rubber-stamping. The control point is the moment of acceptance, and that moment must remain real.

The two interactive operating models therefore admit governance at runtime. Someone is in the room when the work happens. Someone signs the work off. The framework can rely, to a meaningful degree, on the discipline of practitioners doing their job properly.

Autonomous Agency changes the picture. The act of governing now happens before the system runs rather than during it. Once an agent is operating within scope, supervising every individual decision is neither feasible nor the point. The decisions that matter are the upstream ones: the objective the agent pursues, the guardrails that bound it, the data it can touch, the actions it can take, and the conditions under which it must hand back to a human.

Accountability for an agent's behaviour therefore flows to whoever specified its objective and approved its scope. That is rarely a single individual; it is typically a designated owner supported by risk, legal and technology functions acting as a second line. The discipline closely resembles model risk management as practised in regulated finance. Documented intended use. Validated boundaries. Monitored performance. A deliberate process for change.

The heaviest governance investment for Autonomous Agency therefore goes into the design phase. Once the agent is live, governance shifts to monitoring, exception handling, and periodic re-review. Organisations that try to govern agents through real-time human-in-the-loop checks usually find they have built neither effective control nor a working agent.

Embedded Intelligence collapses governance even further into the design moment. The intelligence is a feature of a system, not an actor with discretion. Every relevant judgement was made when the system was designed, tested, and shipped. Accountability flows to the product owner and the engineering function that approved the release. There is no runtime governance to perform, only design-time governance to have done well, and the design-time work has to be done with the seriousness of a control, not the casualness of a feature.

The implication is structural. An organisation running all four operating models needs governance machinery that operates at runtime for some workflows and at design time for others, and that recognises which is which. A single uniform framework cannot do both jobs.

27.3 The three-tier control framework: approve, decide, automate

Translating these accountability patterns into operations requires a control framework that explicitly classifies decisions. A three-tier classification is the most useful structure available. Decisions humans must approve. Decisions AI may make within defined scope. Decisions that are fully automated by design. Every workflow that touches AI should have its decisions allocated across these tiers deliberately, and the allocation should be visible to the people doing the work.

Decisions humans must approve are those where the consequences of error are severe, irreversible, or carry obligations the organisation cannot delegate. Approving a credit facility above a threshold. Signing a clinical treatment plan. Releasing a public statement under the firm's name. Authorising a payment. The AI may do most of the analytic work; the human must accept it, knowingly, with their name on the decision. Approval is not a checkbox. It is the act of taking responsibility for a recommendation produced by the partnership.

Decisions AI may make within scope are the tier most organisations get wrong. The other two are at least intuitive: a human approves the high-stakes decision, and an engineered system runs the high-volume one. The middle tier is where judgement is genuinely shared with a non-human, and that is uncomfortable enough that organisations tend either to collapse it back into the approval tier (defeating the point) or to wave at it with a vague phrase like "AI may handle routine matters" (defeating the control).

The tier requires real specification. The cost of error must be bounded, meaning a single mistake does not produce an unrecoverable outcome and a sequence of mistakes is detectable in time to intervene. The population of cases must be well understood, which is harder than it sounds: a workflow that looks routine in aggregate often contains a long tail of edge cases that the agent will encounter and that no one has thought through. The boundaries must be testable, meaning someone can construct a case that should fall outside scope and verify that the system refuses it or escalates.

Concrete examples make the tier livable.

Routing a customer enquiry to the right team. Drafting a routine response for later sending. Scheduling a follow-up. Selecting the appropriate checklist for a workflow. Updating a non-material field on a record. None of these are governance-free. Each requires a defined scope, a defined exception path, and a sampling regime that confirms the agent is operating inside the lines and surfaces drift before it becomes a pattern. The control here is the scope, and the scope is only as good as the discipline applied to defining and revisiting it.

Two failure modes recur. The first is scope creep, in which an agent originally permitted a narrow set of actions accumulates additional ones over time without those additions passing through the same review the original scope did. The second is undefined exception handling, in which the scope describes what the agent may do but is silent on what happens at the edge, leaving the agent to improvise in cases for which improvisation is exactly the wrong response.

Fully automated decisions are those engineered into systems where the design is the decision. Fraud screening. Document classification. Anti-money-laundering filters. The control point is upstream, at the design and the change process. Governance is exercised through model validation, monitoring, and recalibration, not through individual approvals. Mature organisations have practised this discipline for decades under the heading of model risk; the discipline transfers.

The three tiers map directly onto the operating-model differences described above. Approval-tier decisions sit at the centre of Paired Cognition and Task Delegation, where a named human accepts the work. Within-scope decisions are the natural home of Autonomous Agency, where the agent acts inside guardrails set in advance. Fully automated decisions are the province of Embedded Intelligence, where the design is the control. A framework that treats the tiers as separable from the operating models will eventually misallocate one to the other.

The classification has to be done at the level of the decision, not the workflow. Most workflows turn out to involve more than one tier.

> *A wealth advisory firm puts every workflow it has through the three-tier classification once. Producing a client review pack involves automated data assembly, AI-drafted commentary the adviser must approve, and a covering letter that gets reviewed but not formally approved. The classification is not done at the level of the workflow as a whole. It is done at the level of each decision inside the workflow.*
>
> *Once the firm sees the work this way, the governance question stops being theoretical and starts being practical.*

Doing this allocation deliberately is what turns a vague AI policy into a working control system. It is also what gives practitioners a clear answer to the question they otherwise have to answer privately.

27.4 Governance is the precondition for adoption, not the brake

A persistent misreading of governance is that it slows adoption. The opposite is more often true.

Organisations without governance hesitate, because every individual using AI must privately decide whether their specific use is acceptable, and most reasonable people, faced with that uncertainty, default to caution or to undisclosed use. The first outcome forfeits the value. The second forfeits the value and adds risk. Both are worse than a clear framework that says yes to specific things and no to others.

Clear governance lowers the cognitive cost of every individual decision to use AI. A practitioner who knows precisely which work types permit which models, what data may be uploaded, what review is required, and who to call if something looks wrong does not have to relitigate those questions case by case. They get on with the work.

Adoption accelerates. The friction has moved from the user to the framework, where it can be addressed once and amortised across thousands of subsequent decisions.

Risk-mature industries already understand this. In banking, traders operate inside elaborate limit frameworks not in spite of speed but because of it. The framework removes the question of permission from each transaction, which is precisely what allows the trader to act in the seconds available. AI governance plays exactly this role when it is designed well. It is the precondition for confident, broad-based use, not the obstacle to it.

The reframe matters for an internal reason as well. Governance work that is sold internally as a brake will be resisted, and governance work that is sold as a brake will tend to behave like one. The same machinery, sold as the thing that lets the firm use AI confidently, at scale, with its name on the outputs, will be designed differently and received differently. Words shape the artefact.

27.5 Governance structures vary with organisational scale

What governance looks like in practice varies with scale, and the variation is substantial.

At the team level, governance can be lightweight and largely embedded in the team's own knowledge base. A short charter describing which operating models the team uses, which workflows are in scope, what data may be used, and how outputs are reviewed is often enough. The team lead is the accountable owner. There is no committee. There is a working agreement and an open conversation when something looks unusual.

At the business unit level, governance becomes a defined function. A nominated AI lead, supported by representatives from risk, legal, technology and the business, runs a forum that approves new use cases, reviews incidents, maintains the inventory, and connects to enterprise policy. The cadence is typically monthly, with a tighter rhythm during early deployment.

At the enterprise level, governance becomes a layered system. A senior committee, often chaired by a chief risk officer or a chief operating officer with executive support, sets policy and risk appetite. A central function maintains the inventory, the standards and the assurance programme. The business units operate within those standards. Internal audit independently assures the whole. The shape mirrors mature operational risk governance because the underlying problem is the same.

The mistake to avoid is applying enterprise machinery to a team-level question, or relying on a team-level charter where enterprise machinery is required. Both are forms of mismatched governance, and both produce friction without producing control.

27.6 Four ways governance fails

> *Stanford HAI's 2025* AI Index, *drawing on the AI Incident Database, recorded 233 reported AI incidents in 2024, a 56 per cent year-on-year rise and a record high. The 2025 figure was 362 (Stanford Institute for Human-Centered Artificial Intelligence 2025). The MIT AI Incident Tracker now classifies more than 1,400 real-world incidents. The volume is not the point. The trajectory is.*

Four failure patterns recur, but they are not equally common and they are not equally damaging.

The simplest is having no framework at all, leaving each user to improvise. This is the failure of organisations that have not yet started, and it usually corrects itself the first time something goes publicly wrong. IBM's 2025 *Cost of a Data Breach Report* found that 63 per cent of organisations had no AI governance policies in place, and that 97 per cent of organisations which had experienced an AI-related security incident lacked proper AI access controls (IBM Security and Ponemon Institute 2025). The gap between AI use and governance is the proximate cause of measurable harm, not a theoretical risk. Overcorrection is the next most visible: governance so restrictive that no realistic use case clears it, and the organisation forfeits the benefits while still carrying the costs of shadow use.

It is painful, but it tends to be self-diagnosing because the people who would have used AI are vocal about why they cannot. A third pattern is monolithic design, in which a single framework calibrated to one operating model is applied indiscriminately across all four. This is the structural mistake the chapter has already named. It resembles competence, which is what makes it durable, but it can usually be unwound once someone identifies it.

The fourth pattern is the one that does the most damage, and it is by some distance the most common in organisations that have invested seriously in their governance programme. It is governance as documentation. The policies have been written. The committee has been chartered. The inventory exists, somewhere, in a SharePoint site that three people have edited in the last six months. The framework, on paper, is complete. Look closer and the flow has stopped. Use cases proceed without ever being tabled. Incidents are quietly handled inside the team that experienced them and never enter the inventory. The committee meets, but the agenda is mostly status updates rather than live decisions. The audit trail looks impressive at the level of artefacts and is empty at the level of choices.

This failure is uniquely costly because it produces every appearance of compliance. Regulators see a framework. Executives see a charter. Auditors see policies. The organisation believes itself governed and behaves as if it were, which means it takes risks under the assumption of oversight that is not actually occurring. When something fails, the failure is not a small one, because the conditions for catching it earlier were nominally in place and were not, in fact, operating.

It is also the hardest to detect from the inside. The other three failures hurt visibly. This one is comfortable. The people who would notice are typically the people whose role is to maintain the documentation, and the documentation is fine.

The corrective in each case rests on the same principle: that governance is a working decision-making system rather than an artefact, and its health is measured by whether real questions are arriving at it, real choices are being made through it, and real changes are being adopted as a result. When the flow stops, the framework is dead regardless of how thick the binder.

A governance framework that is alive will look noisy. New use cases arrive. Some are approved. Some are sent back. Incidents are discussed openly. The inventory changes. People in the business know who to call. None of this looks like a tidy artefact. All of it looks like work being done.

27.7 Governance lives in the knowledge base or it does not live at all

A governance framework that exists only in policy documents is fragile. A framework that lives in the organisation's working knowledge base is durable. Governance decisions, the operating model chosen for each workflow, the scope granted to each agent, the data classifications, the escalation routes and the incident learnings all belong in the same shared substrate that the rest of the cognitive partnership relies on.

This integration is more than convenience. It is the mechanism by which governance reaches the point of work. When a practitioner opens a workflow and the knowledge base presents the relevant operating model, the applicable controls, and the recent incident notes, governance becomes a property of the work itself rather than an external check. The framework is consulted because it is in the room.

Organisations that take this seriously find that governance and capability stop being in tension. The knowledge base captures decisions, the governance layer shapes them, the operating models execute them, and the people are supported throughout. Oversight, accountability, and control turn out, on close inspection, to be the same instruments that good organisations have always used to do serious work, now applied to a new kind of partner.

The senior associate who opened this chapter does not need to read the policy. She opens the matter, and the matter tells her what kind of work it is, which model fits, what is permitted, what must be approved, and what to do if something looks wrong. She gets on with the work.

That is what governance is for.

Chapter 28

Why AI Adoption Fails, and How to Make It Stick

> *There is nothing more difficult to take in hand, more perilous to conduct, or more uncertain in its success, than to take the lead in the introduction of a new order of things.*
>
> — Niccolo Machiavelli

Most AI programmes do not fail because the models are inadequate.

Governance permits the work. It does not, on its own, ensure the work is well chosen. Programmes fail because the organisation never decided what humans and AI would do together, and a year of activity has accumulated around that unmade decision. The failures happen inside the boundary governance has drawn, where the framework that permits the work cannot see whether the work itself has been designed.

The pattern is the same across finance, legal, healthcare, insurance and technology. Capable models are procured. Licences are distributed. Announcements are made. Twelve months later, usage is patchy, value is unclear, and a second wave of investment is being justified to the board with the same slides as the first. Diagnosed honestly, the failures cluster into a few recurring patterns that feed into each other. None of them are problems of model quality.

What follows traces the failures from where they originate, upstream of the work, down through training and governance, and out to the ways the programme presents itself to the people who have to live inside it. The sequence is also the prescription, because each failure mode inverts into the discipline that makes adoption stick.

28.1 A procurement decision that was never an operating decision

The most common failure is also the least visible. The organisation has not chosen an operating model. It has procured a tool and assumed an operating model will emerge from contact with the work. None does. What emerges instead is whatever each individual practitioner improvises, which means the organisation is running four or five informal models in parallel without naming any of them.

Paired Cognition, Task Delegation, Autonomous Agency, and Embedded Intelligence are not interchangeable. They differ on the axis of decision-making responsibility, and that axis is load-bearing. Choose wrong, and the work either grinds against the model or escapes it entirely.

> *The regional bank introduced earlier in the book applied Task Delegation to credit memo drafting, and analysts received cleaner first drafts measurably faster.*
>
> *The same bank then applied Task Delegation to restructuring decisions on stressed exposures. The outputs were technically correct and contextually wrong. The work required Paired Cognition: a shared mental model of the borrower, the covenant package, the workout history, the relationship overlay. Cold briefs do not produce that. The programme was judged a failure of AI when it was a failure of model selection.*

The mirror image appears at the other end of the spectrum. An insurer applies Paired Cognition to high-volume, low-variance adjudication that should sit inside Embedded Intelligence. Adjudicators end up reviewing every output. Throughput collapses. The programme is shut down and the organisation concludes AI is not ready for claims, when in fact the wrong model was placed against the work.

The same root cause produces a related failure: starting with the tool rather than the workflow. A legal team rolls out a contract review assistant because the vendor is convincing, then asks which matters it should be used on. A technology firm adopts a coding assistant company-wide and measures success by lines of generated code accepted. In neither case has the workflow been redesigned around any operating model, which leaves the tool sitting beside existing practice rather than inside it. Licence utilisation rises while output quality stays where it was. Senior practitioners quietly route important work around the system.

The two failures are versions of the same mistake at different scales. One skips the model selection question; the other skips the workflow redesign question. Both confuse buying capability with designing cognition. Buying is fast. Designing is slow. Programmes that confuse the two end up paying for the first and never doing the second.

The diagnostic question developed earlier in the selection discipline is worth restating here. Not which tool, not which vendor, but which operating model fits the cognitive demands of this specific workflow.

28.2 When training teaches the interface, what is missed?

Training failures fall into two categories. The first is the absence of training, which still occurs more often than is admitted. The second, more insidious, is training that teaches the interface rather than the partnership.

Staff learn how to prompt, how to attach files, how to use chat history. They do not learn how to build a shared mental model with the system, how to recognise when an answer is plausible but unsafe, how to handle the Reflection phase, or when to escalate from Paired Cognition to a human-only decision. The training looks rigorous and produces almost no durable practice.

> *An insurer trained 500 underwriters on prompt patterns over two days. Enthusiasm was high. Within a quarter, usage had reverted to a small minority of early adopters, and the rest had quietly returned to pre-rollout practice.*
>
> *A different insurer, in a similar segment, trained 200 underwriters on jointly reasoning through a risk with the system, on capturing decisions back into a shared knowledge base, and on what the Reflection phase looked like at the end of a complex submission. Twelve months later, retention of practice was high, the knowledge base had grown, and the loss-ratio improvement was measurable.*

What separates the two outcomes is not training intensity but training subject. Interface training produces interface fluency. Partnership training produces partnership.

28.3 Absent governance forces every practitioner to govern alone

Governance is often treated as a brake on adoption. In practice, its absence is one of the surest ways to stall adoption. Without clear rules on which model applies to which workflow, what data may be used, how outputs are reviewed, and how incidents are handled, individual practitioners must make those judgements alone. Some will be cautious and underuse the system. Others will be incautious and create exposures. Both outcomes erode trust.

> *In a global law firm, partners who lacked written guidance on AI use in privileged matters defaulted to forbidding it entirely. Juniors used personal accounts on commodity models to get the work done. The organisation acquired risk and forfeited benefit simultaneously. A short, explicit governance document, written and circulated within a fortnight, removed both behaviours. Adoption increased. So did the firm's risk posture.*

Clear governance, paradoxically, accelerates adoption. It lowers the cognitive cost of every individual decision to use the system. The practitioner does not have to reason from first principles each time about whether this matter, this data, this output is permitted. The reasoning has been done once, centrally, and is available as a default.

The previous chapter develops the governance framework in detail. The point here is narrower. The absence of governance is itself a failure mode, not a neutral starting position.

28.4 Theatre at the top, replacement at the bottom

The remaining failures share a feature: they are about how the programme presents itself rather than what it does. At the executive level, this looks like transformation theatre. At the practitioner level, it looks like a replacement narrative. Both produce the same result, which is a programme that consumes investment without changing how decisions get made. Prosci's longitudinal *Best*

Practices in Change Management benchmark, now in its eleventh edition and consistent across nine consecutive studies since 1998, finds that projects with extremely effective sponsors are 79 per cent likely to meet their objectives, against 27 per cent for projects with extremely ineffective sponsors; Prosci also reports that roughly half of executives do not understand their sponsorship role (Prosci 2024). Theatre is what fills the gap when sponsorship is rehearsed rather than understood.

Transformation theatre is the production of visible activity around AI without the underlying operational design that would make AI valuable. Workshops happen. Frameworks are introduced. Executives become fluent in the vocabulary. Centres of Excellence are stood up. Strategy documents accumulate. Vendor demonstrations multiply.

Meanwhile, in the workflows where decisions are actually made, the relationship between humans and AI remains undefined. Pilots run. Results are mixed. Conclusions are unclear. The cycle repeats.

The missing element is operational design at the level of specific workflows. A healthcare network may have run three Centres of Excellence and produced two strategy documents and still have no defined Paired Cognition workflow for any clinical or operational task. The activity is real, and the change is not. The diagnostic test is simple: name a workflow, name its operating model, name the seven phases as they apply, and name where Reflection is captured. If those questions cannot be answered for any specific workflow, the programme is theatre regardless of how much activity surrounds it.

The replacement narrative is theatre's mirror at the other end of the org chart. When AI is introduced as a way to do more with fewer people, teams correctly infer that they are the cost being optimised. Resistance follows, sometimes overt, more often quiet. Senior practitioners protect their judgement work by keeping it off the system, while junior staff use it without investing in it. The knowledge base never accumulates the institutional reasoning that would have made subsequent work compound. The pattern shows up in the macro data as well. BCG's 2025 *AI at Work* survey reported regular AI use at 72 per cent overall but only 51 per cent among frontline employees, a "silicon ceiling," and noted that the geographies with the highest use, India at 92 per cent and the Middle East at 87 per cent, also reported the greatest fear about automation impact on their roles (Boston Consulting Group 2025a). High usage and high anxiety travel together when the framing is replacement.

This is where the operating model choice has its most powerful organisational consequence. Automation-first framings, whether Task Delegation pushed past its useful range or premature Autonomous Agency, signal replacement. Paired Cognition signals augmentation. The difference is felt long before it is articulated. A workforce that believes the project will eliminate them does not contribute the externalised expertise the system needs to become useful. A workforce that believes the project will extend them does. The difference between those two outcomes is rarely the technology.

28.5 The resistance paradox

Here is a claim that runs against almost every assumption an adoption programme is built on. Of the four operating models, the one that tends to produce the least organisational resistance is Paired Cognition, the model most people expect to produce the most.

The intuition runs the other way. Partnership sounds slower, more intimate, more demanding. Surely teams would prefer to keep AI at arm's length, delegating discrete tasks rather than thinking alongside it. Surely the deeper the integration, the harder the social uptake.

The intuition is wrong. The reason is human before it is technical. Teams resist AI when they experience it as a threat. They embrace it when they experience it as a partner. The other three operating models, in different ways, present the system as something that takes work away. Paired Cognition presents it as something that thinks alongside. The first reading triggers self-protection. The second triggers curiosity. Curiosity scales. Self-protection does not.

The mechanism is recognisable on the first session. The system holds context with the practitioner, surfaces considerations they had not reached yet, executes within a shared mental model rather than from a cold brief, and learns alongside them through the Reflection phase. Practitioners who walked in expecting to be made redundant walk out having had a better afternoon's thinking than they have had in months. The experience is hard to unlearn. It is almost impossible to argue someone out of afterwards.

So the paradox is structural, not incidental. The operating model that takes the people most seriously is also the one that adopts the fastest. Adoption becomes something practitioners drive for themselves rather than something imposed from above, because they are the ones who felt the difference first.

> *A wealth advisory team began with two advisors using Paired Cognition for client review preparation. The pilot was small, deliberate, and unmandated. Within 10 months, the practice had spread firm-wide, not because of a rollout plan, but because the advisors who had it would not give it up and the advisors who did not have it asked for it. The firm's adoption budget for the year was largely unspent. The adoption had funded itself through demand.*

A similar pattern has been observed in settings that look almost nothing like a wealth firm. In a regional public health service, a multidisciplinary team responsible for complex case planning began using Paired Cognition to think through care plans for patients whose presentations cut across specialties. The team had been chosen for an unrelated pilot and was, by reputation, sceptical: senior clinicians, regulated practice, an institutional culture that treats new tools with appropriate caution. The first sessions did not feel like adoption. They felt like a slightly more structured case conference, in which the system held the history, surfaced considerations the team had not yet reached, and stayed inside the clinical reasoning rather than around it. Within a year, neighbouring teams had asked to join, the executive sponsor's main task had become managing demand rather than generating it, and the framing inside the service had shifted from whether AI belonged in clinical work to which workflows would adopt it next. The shape of the organisation was different. The direction of pull was the same.

Most adoption programmes get this exactly backwards. They build a central engine to push the practice outward against expected resistance. The right operating model inverts the direction of pull. Adoption arrives from the periphery, and the centre's job is to remove obstacles fast enough to keep up.

28.6 Successful adoption is specific, embedded, and compounding

Durable adoption shares a structural signature. It is specific, embedded, and compounding. Each word is doing work.

Specific means it is defined at the level of a workflow, not a department. The credit committee pre-read. The discharge summary. The policy renewal review. The post-incident report. Adoption does not happen to "the legal team" or "the underwriting function." It happens to a named piece of work with a named operating model.

Embedded means the operating model is part of how the work is done, not a parallel track that runs alongside it. If the practitioner can do the work the old way and the new way, and chooses each time, adoption has not happened. The new way is the way.

Compounding means the knowledge base grows with use. Each engagement leaves something behind that makes the next engagement faster, sharper, or safer. This is the engine that turns AI from a tool into an organisational asset.

The Reflection phase is the mechanism of compounding. It is distinct from Completion, which closes administrative formalities. Reflection captures decisions worth recording, documentation to update, knowledge to add to the base, lessons learned, workflow improvements and who needs to be informed. Skipping it severs the feedback loop that makes the model improve in value over time.

Over months, the knowledge base reduces reliance on individual domain knowledge. New hires reach competence faster because the institutional reasoning is available to them. Continuity is protected when senior staff are unavailable. The shared mental model lives in something more durable than a single person's head.

This is the prescription. Specific, embedded, compounding. A programme that lacks any of the three will stall.

28.7 Adoption travels through three phases and most stalls happen at one seam

Adoption typically passes through three phases.

In the pilot phase, a small number of workflows are redesigned with an explicit operating model and a small group practises the new method, with Reflection captured deliberately. In the embedding phase, the redesigned workflows become the default for that work, governance is formalised and training is extended. In the compounding phase, the knowledge base becomes a strategic asset, new workflows are added at lower cost because the patterns are known, and the organisation begins to get better at adopting.

Most stalls happen between pilot and embedding. The pilot succeeds. The team that ran it is celebrated. The practice fails to cross into adjacent workflows because the operating model was never made explicit enough to be portable. A fintech that ran an excellent Paired Cognition pilot in fraud analytics could not replicate it in compliance because no one had written down what made it work.

The corrective is unglamorous. Write down the operating model. Write down the seven phases as they applied. Write down what Reflection captured. What carries forward from a pilot is the documented pattern it produced, not the pilot itself.

28.8 Leading indicators distinguish real adoption from surface activity

Although usage is often mistaken for adoption, the leading indicators of genuine adoption are different and worth naming.

Practitioners can articulate which operating model applies to which workflow and why. The Reflection phase is being completed and the knowledge base is growing. New hires reach competence faster than they did a year earlier. Senior practitioners route important work through the system rather than around it. Governance questions are being asked early in workflow design rather than retrofitted afterwards. Decisions about new workflows reference the existing knowledge base rather than starting from scratch.

Surface adoption shows the opposite signature. High licence utilisation, low knowledge base growth, no shared vocabulary for operating models, and senior staff who use the system for low-stakes tasks while keeping the consequential work off it.

Most programmes actually live in the gap between those two signatures, accumulating usage without accumulating capability rather than failing in any visible way.

28.9 When adoption stalls, the productive move is diagnostic

When adoption stalls, the instinct is to procure more capability or run more training. The more productive move is diagnostic.

Which workflows have an explicit operating model, and which do not? Where is Reflection being skipped, and what has the knowledge base failed to learn as a result? Where has Task Delegation been pushed into work that requires Paired Cognition, or Paired Cognition imposed on work that should be Embedded Intelligence? Where has governance been left implicit, forcing every practitioner to make the same judgement calls alone? Where has the replacement narrative leaked into how the programme is described, and what would it take to re-frame it?

In almost every case, the corrective action is to return to a single workflow, define the operating model precisely, restore the Reflection phase, and let the practice spread from there. Organisations that adopt well do so one workflow at a time, with the operating model chosen deliberately. The compounding takes care of the rest.

The failure modes catalogued in this chapter are recognisable across every industry the book has covered. So is the prescription. The next chapter turns from the structural conditions for adoption to the human ones: what changes inside a team when partnership becomes the way the work is done.

Chapter 29

The Human Side of AI Partnership: What Changes When Teams Work With AI

Making is thinking.

— Richard Sennett

A senior consultant in a Sydney advisory firm, twelve weeks into her team's adoption of Paired Cognition, said something her partners did not expect. She said the work felt lighter. Not easier. The problems were as hard as ever, the clients as demanding, the deadlines as tight. Lighter in a different sense. She was no longer carrying every thread alone in her head between meetings, no longer waking at four in the morning to keep the model intact until daybreak. Something had changed in the texture of the working week.

That kind of remark is the real test of whether an operating model has embedded. The framework can be drawn on a whiteboard. The seven phases can be memorised. The conversation types can be itemised. None of that tells anyone whether the organisation has actually shifted. The shift shows up in how professionals describe their own work, and in what they no longer say.

This chapter looks at the human layer of that shift: what changes inside teams when Paired Cognition takes hold, what stays the same, and what leaders have to do, quietly and by daily example, to make the change real rather than announced.

29.1 A new rhythm replaces the silent carrying of thought

In traditional knowledge work, an enormous amount of thinking happens silently. A partner turning over a client problem on the commute. An associate mentally drafting an argument in the shower. A controller working through a variance in the half hour before the household wakes. The thinking is real and often the highest-value thing the professional does all day. It is also invisible to colleagues, durable nowhere and impossible to share.

Paired Cognition changes the location of that thought. Reasoning that previously lived only in someone's head now happens in dialogue. Ideas surface earlier. Half-formed objections get tested before they harden into late surprises. The work accumulates as artefacts the team can return to, rather than evaporating with the end of the day.

The rhythm of a project shifts accordingly. It feels less like sprinting between deliverables and more like sustained pace. Meetings change with it. They become richer because more of the thinking has already happened before anyone arrives in the room.

> *A strategy team preparing for a client steering committee used to spend the first thirty minutes of each internal meeting catching up on where everyone's thinking had drifted since the last conversation. After three months of Paired Cognition practice, those thirty minutes disappeared.*
>
> *People came in with structured drafts, considered options and pre-explored objections. The meeting itself moved straight to the questions that genuinely required collective judgement. The Friday five o'clock meeting that used to run to seven now ended at six.*

Over time, a shared mental model develops not just between the human and the AI, but across the team. The knowledge base grows. Patterns get recognised and named. Decisions and the reasoning behind them are captured in places people can find them. Work feels more supported because the support is genuinely there, distributed across human, AI and the accumulated record of how the team thinks together. The three-node architecture introduced earlier, of human, AI, and knowledge base, stops being a diagram and becomes a felt reality.

29.2 The worry beneath the worry is about identity, not employment

When professionals first hear that their organisation is bringing AI into their work, the surface conversation is about jobs. The deeper conversation is about identity. A lawyer who has spent twenty years learning how to read a contract, a clinician who has spent fifteen learning how to read a patient, an underwriter who has spent ten learning how to read a submission, are being asked to imagine a world in which that expensively earned reading might matter less. The question is rarely just economic. It is about whether the work that has shaped a professional life will continue to mean what it has meant.

This is the worry beneath the worry. It deserves to be named directly, because it is real, and because most adoption programmes never have the conversation in which it could be answered. Pew Research Center's February 2025 survey of 5,273 employed US adults reported that 52 per cent of workers were worried about future AI use in the workplace, 33 per cent felt overwhelmed and only 36 per cent felt hopeful, and just 6 per cent expected AI to lead to more job opportunities for them in the long term, against 32 per cent who expected fewer (Pew Research Center 2025). The numbers describe a workforce that has already done the emotional arithmetic, and the conclusion is not optimism.

Automation-first framing makes the worry worse, not better. When AI is presented as a way to remove human effort, professionals correctly hear that their effort is the thing being targeted. Even when the framing is softened with language about freeing people for higher-value work, the implicit hierarchy is unchanged. The human is cast as the bottleneck and the machine as the solution. The professional, listening carefully, hears that they are now positioned as a cost.

Paired Cognition reframes the relationship at its foundation. Rather than functioning as the bottleneck, the human becomes the source of judgement, context, accountability, and care. The AI is a partner that extends the reach of those uniquely human contributions. Professionals encountering this framing for the first time often visibly relax, because it matches their lived experience of what makes their work valuable. They have always known that their judgement was the point. They had begun to wonder whether the organisation knew it too.

> *A regional law firm running an internal session on AI started the morning with a conventional automation pitch. The room was guarded. Arms crossed, questions sharp, the particular quiet of senior practitioners deciding to wait this one out.*
>
> *After the break, the framing changed. AI as a partner that lets a lawyer think with more reference material in view, hold more parts of a brief at once, draft closer to the lawyer's own voice rather than further from it. By the end of the day, the same partners who had been sceptical in the morning were asking how soon they could pilot it on their next matter. Nothing about the technology had changed. The framing of the human had.*

This is the resistance paradox named in the previous chapter, and it is worth restating in human terms here. Teams that were sceptical about AI tend to become enthusiastic when they experience it as a cognitive partner rather than a replacement. Adoption stops being something imposed from above. It becomes something people drive for themselves, because the experience of working this way is genuinely better than the alternative.

29.3 What does not change?

It is worth being precise about what AI partnership leaves untouched. The list is longer than most adoption programmes acknowledge, and naming it explicitly does more for trust than any reassurance about employment ever could.

Judgement remains human. The decision about what matters in a particular situation, which client is ready for which conversation, which risk is worth taking and which is not, which inconvenient finding has to be raised even though it complicates the deal, continues to belong to the professional whose name sits on the work. AI can surface considerations. The weighting of those considerations against each other is judgement, and judgement does not transfer.

Domain expertise remains human. The lawyer who has tried fifty cases knows things about how a courtroom actually moves that no model can reconstruct from training data alone. The underwriter who has assessed a hundred property submissions carries pattern recognition built from years of seeing how risks actually behave. The engineer who has watched three plants come online knows which warning signs are real and which are noise. AI can engage with that expertise in dialogue, but it cannot substitute for it.

Creativity remains human in the sense that matters most for professional work. AI can generate options at a scale no human team could match, and the act of recognising which option is genuinely interesting, which one connects to a client's unspoken need, which one breaks a stale frame, is a human act. Creativity in professional work is rarely about producing novelty for its own sake. It is about producing the right novelty for this moment, and that judgement sits with the human partner.

Professional relationships remain human, because clients hire people rather than tools. Trust accumulates between humans across years of shared difficulty, kept promises and recovered mistakes. The lunches, the awkward conversations, the moments where someone says the harder true thing: none of this is delegated, and none of it should be.

Ethical responsibility remains human. Accountability for an outcome, including accountability for how AI was used in producing it, sits with the professional. Regulators understand this. Clients understand this. Professionals understand this in their bones. Naming it openly is part of how leaders demonstrate that the partnership framing is honest, not promotional.

That is a large surface area of what the work has always been, and what the profession was entered to do. AI does not touch any of it. A reader who has worked through Part IV of this book will recognise these constants behind every industry chapter, from the clinician's responsibility for a treatment decision to the litigator's responsibility for what gets filed under their name. The framework rests on this floor.

29.4 What genuinely changes is the surface area a professional can hold at once

What does change is real, and worth being equally precise about.

Coverage improves. A consultant working with AI partnership can engage with more dimensions of a problem in the same amount of time, with more threads staying visible. The result is work that is more thoroughly considered rather than shallower.

Depth of analysis improves because iteration becomes cheap. A first draft can be challenged, restructured, and reframed several times in a session that previously would have produced a single pass. Professionals find themselves exploring second-order questions they would historically have left aside under time pressure. The quality of the resulting reasoning shows it.

Speed of iteration changes the relationship to feedback. Instead of waiting for a colleague to be free to read something, a professional can pressure-test a draft against an AI partner immediately, surface the obvious weaknesses, and arrive at the human conversation with a stronger starting point. Colleagues then spend their time on the questions that genuinely require their judgement, rather than on the easy ones the AI could have caught alone.

Access to broader reference also changes. AI partnership lets professionals draw on adjacent fields, on prior client work captured in the knowledge base, and on structured frameworks they might otherwise have forgotten to apply. The reasoning becomes more grounded. The work feels less like reinventing each time and more like building on a foundation the team owns collectively.

Although some of the work does get faster, the net effect is not speed. The more important effect is that the work gets better in ways that are visible to clients and to the professionals themselves.

29.5 The transition from Task Delegation habits to genuine partnership is uncomfortable before it is rewarding

The shift from Task Delegation habits to genuine Paired Cognition is rarely instant. It has phases, and the middle phase is harder than most adoption plans acknowledge.

Most professionals begin by treating AI as a clever assistant they hand discrete tasks to. They write a polished prompt. They receive an output. They judge it. They move on. The interaction is transactional, the gains are real but modest, and the experience confirms what the professional already half-suspected. AI is a faster typewriter, not a partner.

The transition into partnership begins when professionals start externalising their thinking rather than packaging it. The mechanics matter, because the difference is small on the surface and large in effect. Instead of writing a polished prompt, the practitioner speaks or types their thinking as it actually arises, with its hesitations, its half-finished sentences, its tentative reaches in directions that may or may not pan out. The AI ingests this messy thinking, structures it, and feeds it back. The practitioner sees their own reasoning more clearly than they did when it was only inside their head. This is the externalised cognition loop introduced earlier, encountered now as a daily habit rather than an idea on the page.

In practice it feels distinct from anything taught in professional school. A senior consultant moving from Task Delegation to Paired Cognition might describe the difference like this. Under Task Delegation, the work felt like commissioning. Brief, receive, edit, send. Under Paired Cognition, it feels closer to a long conversation with a colleague who never gets tired and who has read everything in the firm's knowledge base. The thinking happens with the AI, not before it.

Several specific skills have to develop, and naming them helps practitioners locate themselves on the curve. Externalising before structuring, rather than after. Tolerating a half-formed thought long enough to let it land before correcting it. Reading AI output as a draft to be argued with rather than a verdict to accept or reject. Keeping a shared mental model alive across a session, returning the AI to context when it drifts. Closing each session deliberately rather than letting it trail off. None of these are mysterious. All of them cut against habits that twenty years of professional training have reinforced.

This is why it feels uncomfortable. Professionals are trained, often from their first year in practice, to present polished output. Externalising a half-formed thought to anyone, including an AI that no one else will read, can feel exposing in a way that is hard to articulate. The discomfort is not irrational. It is the friction of a long-standing professional habit being asked to do something different.

> *An associate at a mid-tier accounting firm described the first two weeks of trying to externalise as "talking to a wall and feeling stupid." She kept catching herself rewriting her thinking before saying it, exactly as she would have written a brief. Her supervisor asked her to try something specific. At the start of each session, talk for a full minute about the problem before asking the AI for anything.*
>
> *By the third week, she noticed she was hearing herself work things out. By the sixth week, she was running the sessions without thinking about it. By the third month, she said she could not imagine going back.*

That arc, uncomfortable then awkward then natural, is the typical shape of the transition. Knowing the shape in advance helps. It reframes the discomfort as a stage rather than a verdict on the practitioner.

The other skill that develops more slowly is the use of the Reflection phase. Early adopters often skip it, treating each session as a one-off. Mature practitioners deliberately close sessions by capturing what was learned, what worked, what should be carried forward. An earlier chapter makes the case for Reflection at length; this chapter notes only that it is also where the human transition completes. Practitioners who Reflect consistently begin to feel ownership of a body of thinking that compounds. Practitioners who skip Reflection continue to feel they are starting fresh each time, regardless of how good any single session was.

29.6 What leadership has to do is quieter than usual

There is a meaningful difference between announcing AI adoption and enabling it. Announcements happen in town halls and emails. Enablement happens in the quiet daily decisions about what is rewarded, what is permitted, and what leaders themselves are seen doing. Teams read these signals far more accurately than any communication plan assumes.

Leaders who enable Paired Cognition model it openly. They show their own AI partnership in real time, including the messy externalisation, the dead ends, the moments where the AI got something wrong and the leader corrected it. Modelling in pristine form does the opposite of what is intended. It tells the team that mastery is the price of admission. Modelling imperfectly tells the team that the goal is better thinking, not the appearance of effortless command. Gallup's 2026 workplace research found that the strongest single predictor of whether an employee adopts AI is whether their direct manager actively champions it: employees who feel their manager supports AI use are 8.7 times more likely to say AI has transformed their work. Manager engagement itself, Gallup notes, fell from 31 per cent in 2022 to 22 per cent in 2025 (Gallup 2026). The implication for leaders is uncomfortable and clear. The lever that matters most is the daily example of a manager who is themselves visibly working with AI, and that lever is currently weaker than at any point in the last decade.

They also protect the conditions that make partnership possible. They protect time for the Reflection phase that does not produce immediate deliverables. They protect the right to iterate by not punishing rough first drafts. They protect psychological safety for experimentation, which means being explicit that early stumbles with AI partnership are part of building a capability the team will rely on for years.

And they reinforce the framing. When a leader consistently describes AI as a cognitive partner rather than a labour replacement, the team's interpretation of their own work shifts accordingly. People are not being optimised away. They are being given a partner that lets them do the work they entered the profession to do.

29.7 The cultural conditions beneath the leadership behaviours determine whether partnership can take root

Beneath the leader's daily moves sit cultural conditions that decide whether partnership becomes embedded or stays at the level of exhortation.

The first is calibrated trust in AI output. Teams that treat every AI response as authoritative will be burned. Teams that treat every AI response as suspect will never realise the gains. The healthy posture is engaged scepticism, where output is read carefully, challenged where it matters, and accepted where it holds up.

The second is willingness to externalise. This is partly a skill and partly a cultural permission. Teams where colleagues regularly share unfinished thinking with one another find it easy to extend the same posture to AI. Teams where work is only ever shared once polished find externalisation harder, and leaders need to deliberately create the space for it.

The third is comfort with iteration. Paired Cognition assumes the first version will not be the last, and that improvement happens through cycles rather than through a single heroic pass. Cultures that treat iteration as evidence of weakness will struggle. Cultures that treat iteration as the normal shape of good work will thrive.

The fourth is distributed cognition as a working assumption. The team accepts that thinking lives across humans, AI, and the knowledge base together, and that no single node holds the whole picture. This has a quietly humanising effect. Professionals are no longer expected to be always on call, because the knowledge base carries continuity that previously depended on individual availability. Less dependence supports better work-life balance. Better work-life balance supports the sustained attention that good professional work requires.

29.8 What healthy embedding looks like is recognisable from the outside

A team where Paired Cognition has genuinely embedded shows several observable signs. People talk openly about how they used AI in producing a piece of work, including where it helped and where they overrode it. The knowledge base is referenced in conversations rather than ignored. Drafts circulate earlier and in rougher form than they used to.

Meetings are shorter and more decisive because more of the thinking has happened beforehand. New joiners come up to speed faster because the team's reasoning is captured rather than tacit. Senior professionals find themselves engaging with more of the questions they actually entered the profession to engage with, because the routine cognitive load has been redistributed.

Most tellingly, the team's relationship to its own work changes. The pressure of holding everything individually eases. The pace becomes more sustainable. People still work hard. But they work with a partner, and the partner remembers, structures, and extends what they bring to the table.

That is what AI partnership looks like when it is working. It is recognisably different from anything that came before it, and recognisably continuous with the best of what professional practice has always been. The next chapter steps back from the human layer to look at where this practice is heading, and what makes the framework durable as the underlying capability keeps changing.

Chapter 30

The Cognitive Future: Where Paired Cognition Goes Next

> *If we want things to stay as they are, things will have to change.*
>
> — Giuseppe Tomasi di Lampedusa

A framework is only as durable as the principle beneath it. The four operating models presented in this book describe the landscape of human and AI collaboration as it stands today, and the landscape itself is in motion. Tooling is maturing. Model reasoning is improving. The boundaries between the modes will not sit still.

A reader who has worked through the preceding chapters has reasonable grounds to ask what happens next, and what to do about it. The honest answer has two parts. The surface of the framework will change, sometimes substantially. The axis beneath it will not.

This chapter is about the difference between those two layers, and what that difference implies for the organisations now building partnership practice. The argument is not that the future is predictable. It is that a framework anchored in decision-making responsibility holds its shape across the kinds of capability change that are currently visible, and the kinds that are not yet.

30.1 Maturation, not discontinuity

The shape of the next phase is not entirely speculative, because several trajectories are already visible in the existing tooling and the existing framework. Each one is a maturation of something the framework already accommodates rather than a break with it.

Multi-modal capability is expanding the channels through which AI can ingest and produce content. A document, an image, a screen recording and a voice note are increasingly handled inside the same partnership. Real-time collaboration tooling is approaching the point where AI participation in live human dialogue becomes seamless. Reasoning improvements are extending the range of objectives an AI can hold over longer horizons without losing coherence.

None of these trajectories breaks the framework. The chat interface gains voice and vision. The transcript-and-process workflow gains a real-time variant. The well-defined task gains complexity it can hold without fragmenting. Each evolution sits inside a mode the framework already names.

This matters for organisations planning their AI investment. The right question is not which capability will arrive when. The right question is which operating model is best suited to the work, and whether the organisation has built the discipline to use it well. Tooling will close the remaining gaps in time. Discipline takes longer to build, and it is the genuinely scarce resource.

30.2 Observed Conversation evolves into Assisted Conversation rather than being replaced by it

The clearest near-term evolution is the shift from Observed Conversation to Assisted Conversation, the two configurations introduced earlier. The cognitive principle behind both is identical. Multiple humans externalise their thinking through dialogue, and an AI processes that externalised content into structured understanding. The difference is when the processing happens. Observed Conversation handles it afterwards. Assisted Conversation handles it in the room, in real time.

The reason Observed Conversation is the dominant practical approach today is not conceptual. It is mechanical. Real-time transcription is mature, but the orchestration required for an AI to participate usefully in a live human conversation is still under construction. Knowing when to speak, when to retrieve, when to summarise, and when to stay silent is harder than it sounds. Current tooling has not yet integrated those judgements into a seamless experience.

When the tooling matures, the working practice will extend rather than reset. Teams already fluent in Observed Conversation will absorb the real-time variant as an upgrade, not as a transformation.

> *A strategic planning team deciding on a new market entry has spent the last year using Observed Conversation. Their pattern is familiar. They meet, they*

talk, the transcript runs to the AI afterwards, and a structured decision log, an updated assumption register, and a list of open questions are waiting in the team's knowledge base by the time they leave the room.

As Assisted Conversation tooling matures, the same team's working pattern shifts in degree rather than in kind. The decision log appears on a screen as the conversation unfolds. A challenge from the AI surfaces a precedent from a previous engagement halfway through the discussion. The team's habit of externalising thinking and reviewing structured output is the same habit. The latency between speaking and seeing has collapsed.

The teams that will move into Assisted Conversation most easily are those that have already built strong habits around Observed Conversation. They have the discipline of externalising thinking through dialogue. They have the muscle memory of reviewing AI-structured output. They have the knowledge base the real-time AI will draw on. The shift becomes incremental.

Teams without those habits will not be rescued by the new tooling. A real-time AI in the room with an undisciplined meeting produces a faster version of the same incoherence. The capability arrives without bringing the practice with it.

30.3 The boundary between Paired Cognition and Autonomous Agency will keep moving

A second trajectory sits at the other end of the framework. As AI reasoning improves, the scope of well-defined objectives an AI can hold expands. Tasks that today require sustained human judgement at every step will, in time, become tractable as autonomous workflows with the human at the boundary rather than inside the loop.

The pace of this expansion is now measurable. METR's 2025 measure of the human-time-equivalent length of tasks an AI agent can complete with 50 per cent reliability has been doubling roughly every seven months for six years; in 2024 and 2025 the doubling rate accelerated to about every four months (Kwa, West, et al. 2025). Frontier models in early 2025 had a 50 per cent time horizon of around 50 minutes. A horizon that today fits inside a single sitting will, on the trajectory the data describes, stretch into the span of an afternoon's work and then a day's, which is exactly the territory where the boundary between Paired

Cognition and Autonomous Agency starts to move.

It is worth being specific about what this means and what it does not. Work genuinely belongs in Autonomous Agency when the objective is clear, the constraints are well-specified, and the consequences of error are bounded or recoverable. As AI gets better at sustained reasoning, more work meets these conditions. Routine document review at scale, complex but well-bounded research synthesis, and multi-step operational workflows are obvious candidates. So is a growing class of investigative work that today consumes substantial analyst time but does not require partnership-grade judgement at each step.

What does not change is the nature of the work that belongs in Paired Cognition. Decisions where context is ambiguous, where competing values must be weighed, where a relationship is in play, where accountability sits with a named professional remain partnership work regardless of how capable the AI becomes. Where the AI getting smarter would once have implied that the human's judgement was becoming residual, the framework's prediction is the opposite: a more capable AI raises the ceiling on what the partnership can reach.

> *A commercial law firm running a portfolio review agent in 2026 might find, by 2028, that the same agent can absorb several upstream tasks that currently require partnership: an initial diligence pass on a contract repository, a first cut at issues categorisation, and a draft summary by counterparty risk profile.*
>
> *None of this displaces the partnership between the senior lawyer and the AI on the strategic question of which way to litigate, which clauses to negotiate hardest, and what to advise the client. The agent's reach expands. The lawyer's judgement is exercised on a richer base. The boundary moves along an axis that itself stays put.*

Some work will migrate from Paired Cognition to Autonomous Agency over the next several years. That direction of travel is clear; the pace is not, and any specific timetable is guesswork. New work will also enter the Paired Cognition zone, as expanded capability makes previously intractable problems addressable through partnership for the first time. Whether the total volume of partnership work grows or shrinks is genuinely open, but the prediction the framework will not make is the one that says partnership work disappears. The mix will shift, and the category will hold.

The aggregate picture is easier to see than the work-by-work account. The

World Economic Forum's employer survey (World Economic Forum 2025) puts numbers on the rebalancing employers themselves expect:

Task share	Today	2030 expected
Primarily human	47%	~33%
Hybrid (human + machine)	30%	~33%
Primarily technology	22%	~33%

Source: World Economic Forum, Future of Jobs Report 2025.

The hybrid share holds steady and grows slightly, which is the partnership category in another vocabulary. The primarily-human share contracts and the primarily-technology share expands, which is the migration the framework predicts. None of the three collapses to zero. The work that belongs in partnership keeps its weight in the mix even as the boundaries on either side move.

30.4 Where does decision-making responsibility sit, and does it move?

The four operating models differ on a single load-bearing axis: where decision-making responsibility sits. In Paired Cognition, decisions are made jointly through sustained dialogue. In Task Delegation, the human decides and directs while the AI executes against a brief. In Autonomous Agency, the AI decides within boundaries the human has set. In Embedded Intelligence, the AI's contribution is woven into a software product whose behaviour the user accepts as given.

Although this axis is stable across capability evolution, what better models change is not where decision-making responsibility sits but the range of what becomes possible once that responsibility has been allocated. A more capable model in Paired Cognition produces better partnership. A more capable model in Autonomous Agency expands the range of objectives it can pursue. The model itself does not redefine which mode applies to a given piece of work.

This is what gives the framework its forward utility. An organisation that learns to recognise where decision-making responsibility belongs in a given workflow has a way of thinking that survives several generations of capability improvement. The tools change. The diagnostic question stays the same.

Who is accountable for the decision, and what is the right configuration of human and AI given that answer.

Teams that hold this question close will adapt to capability change without panic. Teams that do not will find themselves repeatedly reorganising around whichever tool arrived most recently, mistaking surface motion for strategy.

30.5 The human contributions the framework is built on do not move at all

Beneath the operating model axis sits a deeper layer that does not move with capability. Human judgement remains the source of meaning. The decision about what matters, which problem to solve, which trade-off is acceptable, which client is ready for which conversation, belongs to people. AI engages with these questions without answering them.

Ethical responsibility remains human in the same way. When an AI's contribution shapes an outcome, accountability for that outcome sits with the professional whose name is on the work. This is not a temporary limitation of current models. It is a structural feature of how trust and responsibility operate in professional life. A regulator, a client, an auditor, or a court holds humans accountable, and no shift in capability changes who picks up that phone.

Relationships remain human. Clients hire people they trust, and that trust accumulates between humans across years of conversations and shared difficulty. Teams cohere through shared experience and mutual obligation. AI partnership extends the reach of these relationships without substituting for them.

These constants matter for forward planning, because organisations sometimes assume that as AI capability grows, the human contribution becomes residual. The framework predicts the opposite. As AI extends what is possible, human judgement, ethics, relationship and accountability become more central, not less, because they become the differentiators that capability alone cannot provide. In a market where everyone has access to comparable AI capability, the human contributions are what differ.

30.6 What cannot be predicted, said honestly

Forward-looking writing about AI tends to either over-claim or under-claim. The temptation is to predict the next decade with confidence the writer does not actually have, or to retreat into vague gestures about the future being uncertain.

Since neither extreme serves the practitioner, the honest position sits between them. The trajectories named in this chapter, multi-modal expansion, real-time collaboration, longer-horizon reasoning, are visible, and one of them is now measurable: METR's data on the human-time-equivalent length of tasks frontier agents can complete with 50 per cent reliability shows a doubling roughly every seven months over six years, accelerating to about every four months in 2024 and 2025 (Kwa, West, et al. 2025). A trajectory that consistent over that long a window is the strongest available evidence that the boundary between Paired Cognition and Autonomous Agency will keep moving, and moving faster than most planning cycles assume.

Modes of human-AI work that do not fit cleanly into the four operating models as currently described are likely, not merely possible. The field has produced enough surprise in the last three years to make a fifth mode plausible rather than remote. The framework is offered as a way of thinking that has so far survived the changes the field has produced. It is not offered as a permanent and complete description.

What can be said with more confidence is structural. As long as work is performed inside organisations that bear accountability for outcomes, the question of where decision-making responsibility sits will remain the central question. The configurations available to answer it will keep evolving even as the question itself keeps being asked in the same shape.

That is the basis on which the framework is durable. Not because the surface stays the same, but because the question it answers continues to be the question that matters.

30.7 Three investments compound, but one of them does most of the work

The organisations that adapt fastest as the framework evolves are those investing in three foundations today. None of them depends on predicting which specific capability will arrive when. Each prepares the organisation to absorb whatever does arrive. The three are not equal in difficulty or in consequence, and pretending they are obscures where the real work sits.

The easier of the three is knowledge base maturity. A well-structured, well-maintained knowledge base compounds in value over time and becomes the substrate that future AI capability draws on. Organisations that build this discipline now will have a richer cognitive foundation when more capable tools arrive. The tools work better when the third node is well-tended, regardless of which generation of tooling is current. This is real work, but it is work most organisations already know how to do. It is a knowledge management problem with a clear shape.

Governance is harder, but its shape is also recognisable. The frameworks that work for current AI use will need to extend, not be rebuilt, to handle expanded autonomous agency, real-time assisted conversation, and the wider range of work that future capability will make tractable. Organisations that have already established clear principles for accountability, oversight, audit and incident response will be able to extend those principles to new operating modes without starting from scratch. The ones that have not will find themselves writing governance for the first time under the pressure of an arriving capability. Hard, but tractable. The discipline exists in adjacent domains and transfers.

The third investment is the one this chapter wants to weight most heavily, because it is the one organisations consistently underinvest in and the one that most directly determines whether the other two are useful. Operating model literacy across the workforce is genuinely difficult to build, because it cannot be packaged as a course or a policy document. It comes from the slow accumulation of judgement about which mode of human-AI work fits which kind of problem, distributed across thousands of individual decisions made by people who have to recognise the situation in front of them in real time.

A team with a maturing knowledge base and a strong governance regime, but no operating model literacy, will still misapply Paired Cognition to work that should sit in Task Delegation, or push into Autonomous Agency before the work is well-bounded enough to sustain it. The substrate is there. The discipline is there. The judgement about how to deploy them is not, and the result is expensive AI activity that does not produce partnership outcomes.

Operating model literacy transfers cleanly across capability generations because it is a way of thinking, not a way of using a particular tool. It is also the slowest of the three to build, because it has to be earned through repeated decisions that produce repeated feedback, in the actual work, with real stakes. Knowledge base maturity can be project-managed. Governance can be drafted. Literacy has to be lived into. An organisation that begins this investment three years ahead of a peer is not three years ahead. It is in a different category.

The pattern across the three is the same. Each survives the next generation of tooling because each was never about the tool. The difference is how much of the survival depends on each. The first two prepare the ground. The third is what walks the ground.

30.8 Organisations that master the framework now will adapt faster as it evolves

The argument of this chapter has been building toward a single conclusion that the preceding sections make available rather than assert. If the axis is stable, if the trajectories already visible mature inside the framework rather than break it, and if the three foundational investments survive any specific capability shift, then the organisations that build partnership discipline today are not preparing for one future. They are preparing for the family of futures the axis allows.

The advantage compounds. A team fluent in Observed Conversation will adopt Assisted Conversation as an extension of practice they already have, not as a new skill to teach from scratch. A team disciplined about where decision-making responsibility sits will incorporate expanded Autonomous Agency without losing track of accountability when the boundary moves. A team with a maturing knowledge base will see each new capability extract more value than the same capability would extract from a less prepared organisation, because the substrate is already there. Different capabilities, same underlying discipline. The discipline is what compounds.

The opposite case is equally concrete. Organisations that have not built the underlying discipline will continue to struggle as the surface keeps changing. New capabilities will not rescue them. They will add new ways to use AI badly, faster, and at greater scale. The framework will not have failed. The investment in the framework will not have been made.

This is the closing argument of the book, and it is that the framework is not a snapshot of a moment in AI history but a way of thinking that survives the moment because it rests on a stable axis: who is accountable for the decision, and what is the right configuration of human and AI given that answer. The work for any organisation reading this is not to wait for clarity about what comes next. It is to build the discipline that makes whatever comes next genuinely useful when it arrives, on the working assumption that the tools available eighteen months from now will reward the practice an organisation has built today and punish the absence of it. The next chapter takes that assumption and lays out what to do with it on Monday morning.

Chapter 31

What to Do Now: Immediate Actions for Organisations

> *How we spend our days is, of course, how we spend our lives. What we do with this hour, and that one, is what we are doing.*
>
> — Annie Dillard

Every reader of a book like this eventually arrives at the same question. The framework is now visible. The operating models, the work phases, the knowledge base practice, the governance design, the orchestration discipline, the cultural shifts have each been examined in their place. What is missing is the small and decisive matter of what to do on Monday morning.

This chapter is the answer.

It is not a summary. The point of an action agenda is to make the next step obvious, and the next step is not the same for every reader. An organisation that has barely begun needs different work from one that has been running pilots for a year, and both need different work from one that has already embedded several models across its operations. A useful closing chapter has to meet each reader where they actually are.

What follows is shaped accordingly. Three forward-facing leadership prompts open the chapter, then priorities specific to where the organisation already is, the shortcuts most worth refusing, a ninety-day plan and a longer arc into the phase where adoption begins to compound. The aim throughout is specificity. Vague programmes produce vague results, and an organisation that has finished this book has earned a more concrete place to start.

31.1 Three prompts a leader should already be acting on

Every organisation pursuing AI is now past the point where diagnostic self-examination is the work. The work is forward motion, and three prompts shape the next move. Which workflow does the organisation move into Paired Cognition next quarter, and what changes in how the team operates when it does. Which institutional reasoning, currently held in three or four senior heads, is captured into the knowledge base before the next senior departure. Which named individual carries accountability for AI decisions in each business area, with the authority to decide, escalate, or refuse, by a date already on the calendar.

These are not questions about whether the work has begun. They are questions about what the next concrete move is, with a date and an owner. A leadership team that cannot give those answers within a fortnight is not facing a knowledge problem. It is facing a decision problem, and the decision is theirs.

The first prompt forces operating model selection out of the abstract and into a specific workflow. A general commitment to using AI for productivity does not satisfy it. Naming credit underwriting, complex claims, or strategic planning, choosing the model that fits, and committing to the redesign on a date is what satisfies it. The failure modes catalogued in Chapter 28 are kept at bay by exactly this discipline: tool-first thinking, the replacement narrative, and the comfort of metrics that do not measure what matters all dissolve when a specific workflow has a specific operating model and a specific owner.

The second prompt converts the knowledge base from intent into artefact. Most organisations will point to a Confluence space, a SharePoint folder, or the chat history with a vendor's assistant. None of these is a knowledge base in the sense the framework requires. The institutional reasoning of the organisation has to be captured somewhere it can compound, and someone has to be responsible for tending it. The forward move is to identify which body of senior reasoning is at greatest risk of leaving with its current holder, and to begin the structured capture this quarter rather than after the resignation lands.

The third prompt makes governance operational rather than performative. A committee charter does not amount to accountability. Accountability is a named individual who can decide, escalate, or refuse the use of AI in a specific workflow, and who carries the consequences when the decision goes well or badly. The forward move is to publish those names, by business area, before the next quarter begins, so that every practitioner knows whose call it is and the organisation's posture stops drifting toward whichever cautious or incautious individual happens to choose on a given Tuesday.

Three prompts. Three commitments with dates and owners. The next move follows from the answers.

31.2 If the organisation is at the beginning

For an organisation that has not yet begun in earnest, the order matters. Workflow mapping comes first, because everything downstream depends on it. Operating model selection then follows, applied to two or three workflows that genuinely matter. Pilot design comes last. Skip any of these and the pilot cannot be evaluated, because no one is quite sure what was being tested.

Workflow mapping means identifying the specific pieces of work where AI could genuinely change outcomes, and describing those workflows precisely enough that the cognitive demands are visible. The work is rarely glamorous. It is sitting next to a credit officer for a morning, or a claims handler, or a senior associate, and watching what they actually do, including the silent pauses where the real reasoning happens. The discipline is developed earlier in the book. Without it, every later choice is guesswork.

Operating model selection then matches each mapped workflow to the model whose cognitive shape fits, against the six selection criteria established earlier. High-volume, low-variance, well-bounded work tends toward Embedded Intelligence or Task Delegation. Judgement-heavy, context-dependent work where outcomes matter and reasoning needs to be jointly held tends toward Paired Cognition. Narrow, well-defined objectives with stable rules can move toward Autonomous Agency. The selection is per workflow, never per organisation.

Pilot design then closes the loop. A pilot is a deliberately scoped trial of a redesigned workflow with the operating model made explicit, the Reflection phase included from the outset, and success criteria agreed in advance. An earlier chapter on piloting develops this discipline in depth.

> *A regional bank deciding to pilot AI in its credit team begins not with a tool selection but with a single question. What does the credit officer actually do when reviewing a complex commercial application? Watching the work reveals that the officer's value lies in the contextual reasoning that connects financial statements, sector dynamics, and the specifics of the borrower's business. That points unambiguously to Paired Cognition.*

The pilot is then designed around the operating model, not around the software. The tool follows the design rather than driving it.

A pilot is not a vendor demonstration. It is the organisation's first chance to learn how a chosen model works in its own context, and the lessons it generates are the seeds of everything that follows.

31.3 If the organisation is mid-adoption

Organisations already in the middle of adoption tend to have a different cluster of problems. Governance was left implicit. Reflection was assumed to be happening. The knowledge base became whichever shared drive someone gestured at in a meeting, and training stopped at prompting. Each gap was tolerable in the rush to get something working. Each is now beginning to limit how far adoption can go.

Governance gaps appear as inconsistent practice across teams. One business unit pastes confidential client data into a public model. Another refuses to use any AI tool at all because the policy is unclear. Both are symptoms of the same absence. The corrective action is to write the rules down, name the accountable individuals, and make governance a positive enabler of adoption rather than a residual constraint. The three-tier control framework from the governance chapter is the starting point.

Reflection discipline is harder to fix because it requires a cultural intervention. Teams that have learned to celebrate output completion need to learn that the work is not done when the output ships. Decisions worth recording, documentation to update, knowledge to add to the base, lessons learned, workflow improvements: all of these are generated during the Reflection phase. Skipping it severs the feedback loop that makes the operating model compound in value. The Reflection chapter names it as the phase nothing punishes you for skipping, although the damage shows up later, in adoption that does not deepen.

Knowledge base maturity follows naturally once Reflection is happening. The base must be a real artefact, structured well enough that future work can find what past work learned. "It lives on the shared drive" is the answer of an organisation that has not yet treated its institutional memory as a load-bearing asset. Treat it as one. The work then becomes obvious.

Training depth is the difference between staff who know how to prompt and staff who know how to build a shared mental model with a system, recognise when an answer is plausible but unsafe, and know when to escalate to a human-only decision. The training chapter names the four competencies in order. Most stalled adoption programmes have trained for the first and skipped the rest.

> *A professional services firm a year into adoption notices that pilots are producing mixed results despite enthusiastic teams and capable tools. Investigation reveals the pattern. Different practice groups have arrived at different conventions. No one is responsible for the firm-wide knowledge base. Partners assume Reflection happens informally and so it does not happen at all. Training was a half-day session twelve months ago.*
>
> *Each of these is fixable. None of them is solved by a better model.*

The work at this stage is to formalise what was implicit, in the order in which the gaps are limiting growth.

31.4 If the organisation has established practice

Organisations that have already embedded operating models in several workflows face a different challenge. The challenge is orchestration, the work of composing multiple AI methods so they reinforce one another rather than fragmenting.

A well-designed insurer might use Embedded Intelligence in claims triage, Task Delegation in policy administration, Paired Cognition in complex underwriting and Autonomous Agency for narrow reconciliation processes. The disciplines must compose. Composition requires shared vocabulary, shared knowledge base infrastructure, and a clear view of where each operating model hands off to the next. A claim that begins in Embedded Intelligence and escalates into Paired Cognition must arrive at the human partner with context the system has already accumulated, not as a cold brief.

The seams between models are where most orchestration value is lost or gained. The orchestration chapter develops this discipline in detail.

Forward-looking preparation is the second priority at this stage. As the preceding chapter argues, the trajectory of AI capability suggests Assisted Conversation will mature, the boundary of Autonomous Agency will expand, and new operating models may emerge between or beyond the current four. Organisations that have mastered the framework as it stands today will be substantially faster to adapt as it evolves, because the underlying axis of decision-making responsibility is stable even when the technologies change underneath it.

The work to do now is to invest in the assets that survive any evolution of the framework. Knowledge base maturity. Governance literacy. Operating model fluency among practitioners. Tooling will change. The cognitive principles will not.

31.5 Shortcuts that look like progress, and are not

Several shortcuts are tempting at every stage and undermine adoption in predictable ways. Each looks like progress from the outside. Each defers the actual work.

The most common is procuring more capability when adoption stalls. The instinct is reasonable. The action is misdirected. Stalls are almost never caused by insufficient model quality; they are caused by absent operating model selection, missing Reflection, or unclear governance, and adding capability changes none of those things. The new model arrives. The underlying design work is still missing. A year later the same stall is being explained as a need for the next model, and the procurement cycle resumes.

A close cousin is rolling out a tool firm-wide before any workflow has been redesigned around it. This is tool-first thinking. It produces apparent adoption in dashboards while leaving how decisions are actually made unchanged. Senior practitioners quietly route the important work around the tool, send the polished output afterwards, and keep doing the job the way they always did. Usage statistics rise. The programme is declared a success on metrics that do not measure what matters.

Automation-first framing is more dangerous because it poisons what comes after. When AI is introduced as a way to do more with fewer people, teams correctly infer that they are the cost being optimised. Resistance follows, often quietly, and the knowledge base never accumulates the institutional reasoning that would make subsequent work compound. People do not write down what they think the organisation will use to make them redundant. Paired Cognition signals augmentation. As argued earlier, the difference is felt long before it is articulated.

Then there is transformation theatre. Workshops. Frameworks. Executive fluency in the right vocabulary. Centres of Excellence with elegant operating diagrams. None of it substitutes for operational design at the level of specific workflows. An organisation can produce a year of theatre without changing how a single decision is made.

Each of these has the same structure. Visible activity replaces real design. The cure is to return to the three prompts and commit to a date and an owner against each.

31.6 A 90-day plan that puts the foundation in place

A serious organisation can put a meaningful foundation in place in ninety days. The aim is not the full framework, only a foundation strong enough to build the rest on.

The first month is honest stocktaking. A real map of the most important workflows in two or three business areas, and the appointment of accountable individuals for AI decisions within each area. By day thirty, the organisation has named the next workflow to move into Paired Cognition, the senior reasoning at greatest risk of walking out of the door, and the people accountable for AI decisions in each area. That alone places it ahead of most of its peers.

The second month is design. Operating model selection for two or three of the mapped workflows, and pilots designed deliberately to test those selections rather than to demonstrate a tool. The Reflection phase is built in from the start, not retrofitted afterwards. Governance for the pilot scope is written down, and training for the pilot teams covers partnership rather than stopping at prompting.

The third month is when the work actually happens. The pilots run. Reflection is captured systematically rather than as an afterthought. The knowledge base begins to accumulate its first real entries: the decisions made, the conventions adopted, and the moments where the operating model surprised the team in either direction.

> *Day one of a 90-day programme is rarely glamorous. The accountable executive convenes the relevant operational leads, puts the three prompts on the table, and asks for a workflow, a body of senior reasoning, and a name against each, with*

a date next to each commitment. Most of the first answers are tentative. That is the point.

By day thirty the workflow map is on the wall, the operating model selection has been argued through, and someone whose name everyone knows is accountable for AI decisions in each area. By day sixty the pilot is designed, with the operating model named, the Reflection cadence scheduled, and the knowledge base location decided. By day ninety the first cycle of work has produced its first real artefacts. None of this required a new tool that did not already exist.

By day ninety, the organisation has lived experience of at least one operating model in at least one workflow, evidence about what worked and what did not, and the beginnings of an institutional artefact that can be built on. This is a foundation almost no organisation currently has, and it is the start of the work rather than the end of it.

31.7 The longer arc into compounding

Deloitte's late-2025 enterprise survey, published in 2026, reported that only about one in five organisations had a mature governance model for AI, and that the share of pilots moved into full production was still growing slowly, doubling year on year but from a low base (Deloitte 2026b). The implication for the argument here is that a governance build-out which takes twelve to eighteen months is realistic, not pessimistic. Most organisations are nowhere near the finish line, and the ones that are pacing themselves correctly are not behind.

The full framework takes twelve to eighteen months to build out properly. Through the rest of the first year, operating model selection extends across more workflows, governance is formalised rather than improvised, and training deepens beyond prompting into the four competencies named in Chapter 16. Sometime in the second year, the conversation shifts from selection to orchestration: how the operating models compose across the organisation, where the seams between them sit, and what the knowledge base needs to look like to serve all of them at once.

By the end of this period, adoption has crossed from pilot, through embedding, into the early phases of compounding. The phase shift is unmistakable when it happens.

New workflows can be added at lower cost because the patterns are known. New hires reach competence faster than they did a year earlier, because the knowledge base now carries reasoning the organisation used to keep in individual heads. A senior leaver no longer takes irreplaceable context with them. The organisation gets better at adopting, and the gap between it and slower-moving peers widens with every month.

The scale of the asymmetry, once it has had time to compound, is now visible in cross-industry data. BCG's 2025 study of large enterprises separated a small group of "future-built" organisations from the long tail of laggards and reported the following spread.

Metric	Future-built (5%)	Laggards (60%)
Revenue growth	1.7×	baseline
Three-year TSR	3.6×	baseline
EBIT margin	1.6×	baseline
Deeply engaged C-suite	~100%	8%

Source: BCG, Build for the Future 2025.

This is the asymmetry the framework exists to produce. It is not available to the organisation that is still procuring tools and running disconnected pilots in year three. It is available to the organisation that did the underlying design work early and let the discipline compound.

The operating model decisions made today determine AI capability for years. Not because they are hard to revise, but because the discipline that produces a good decision today is the same discipline that produces a good knowledge base, a workable governance posture, and a culture that treats Reflection as part of the work rather than overhead. Organisations that develop that discipline early compound. Organisations that defer it remain in a permanent pilot phase, explaining mixed results to their boards while the next generation of tools arrives to be misapplied in the same way.

31.8 There is no neutral year

The reader who has finished this book is in a position to act, and the action is concrete. Map the workflows that matter, choose the operating model that fits each one and build the Reflection phase into how the work is actually done. Capture what is learned somewhere durable. Name the people who are accountable, refuse the shortcuts and begin where the cognitive demand is real and the outcome is consequential. Let the framework compound from there.

There is no neutral year, because an organisation that defers this work is not thereby standing still. Its competitors are building the discipline, accumulating the knowledge base, and learning to compose the operating models, and each month of postponement widens the gap. The gap is more than usage statistics. It is the depth at which AI is woven into how decisions are actually made: into the reasoning, the institutional memory, the trust calibration and the judgement of when to engage a partner and when to delegate. That depth is not procurable, only built, slowly, by organisations that begin.

The cognitive partnership is not a destination, and it is not a programme to be completed. It is a way of organising work that keeps producing advantage long after the initial decisions have been forgotten. The framework is the scaffolding. The discipline is the substance. Both belong to whoever puts them into practice while everyone else is still procuring.

So the question is not whether the framework works. It does. The question is whether the next twelve months will be spent building the discipline, or explaining to a board why another year of pilots produced mixed results. One of those organisations will compound. The other will keep procuring.

Begin.

About the Author

Carl Andersson is a technologist and AI consultant with 25 years of experience across software engineering, business technology and process design. He has designed AI operating models and orchestration layers for organisations in the Australian financial services and insurance industries, working on the operational decisions that determine whether AI adoption produces real outcomes or expensive disappointment.

He founded Nutropic, an AI strategy consultancy based in Melbourne, to concentrate this work as a dedicated practice. At Nutropic he works with organisations on AI operating model design at the workflow and system level: selecting the right model for each class of work, building the governance and knowledge base infrastructure that makes adoption compound, and ensuring that decisions about cognitive responsibility are made deliberately rather than by default.

Carl is based in Melbourne, Australia.

Acknowledgements

Almost three decades of working in technology means that many of the ideas in this book were formed in conversation, not in isolation. My friend and long-time business partner John Gamble has been instrumental in those conversations, offering me a consistent sounding board and helping cross-pollinate my thinking in ways that are difficult to trace precisely but easy to recognise in hindsight. My thanks to him for that, and for the working relationship that made it possible.

The concepts described in this book were developed in practice, primarily through work with organisations in financial services, insurance and software. My clients cannot be named here, but I am grateful to them for having the courage to act on my advice in domains where the right answer was not obvious, and the established practice did not yet exist. That kind of trust is not given lightly, and it has shaped the framework described in this book more than any other input.

Others have explored similar territory from more formal or academic directions. Hutchins, Clark, Vygotsky and Kirsh all mapped aspects of distributed and extended cognition with rigour that deserves acknowledgement. The convergence between their conclusions and mine is genuine, but it runs in parallel: the ideas here were arrived at through practice, and the theoretical frameworks confirmed rather than produced them.

To Natasha and my family: thank you for your support and patience. I spend a significant amount of time in front of a screen, and your understanding and encouragement make my work much easier than it would otherwise be.

References

Anthropic. 2025a. *Anthropic Economic Index: September 2025 Report*. San Francisco: Anthropic. https://www.anthropic.com/research/anthropic-economic-index-september-2025-report.

Anthropic. 2025b. "Effective Context Engineering for AI Agents." Anthropic Engineering Blog. https://www.anthropic.com/engineering/effective-context-engineering-for-ai-agents.

Anthropic. 2025c. "How We Built Our Multi-Agent Research System." Anthropic Engineering Blog. https://www.anthropic.com/engineering/multi-agent-research-system.

Bank of England and Financial Conduct Authority. 2024. *Artificial Intelligence in UK Financial Services — 2024*. London: Bank of England. https://www.bankofengland.co.uk/report/2024/artificial-intelligence-in-uk-financial-services-2024.

Becker, Joel, et al. 2025. *Measuring the Impact of Early-2025 AI on Experienced Open-Source Developer Productivity*. METR. arXiv:2507.09089. https://arxiv.org/abs/2507.09089.

Boston Consulting Group. 2025a. *AI at Work 2025: Momentum Builds, but Gaps Remain*. Boston: BCG. https://www.bcg.com/publications/2025/ai-at-work-momentum-builds-but-gaps-remain.

Boston Consulting Group. 2025b. *Build for the Future 2025: The Widening AI Value Gap*. Boston: BCG. https://www.bcg.com/publications/2025/are-you-generating-value-from-ai-the-widening-gap.

Cates, John. 2022. "Improving the After Action Review." *Journal of Military Learning* 6 (1). https://www.armyupress.army.mil/Journals/Journal-of-Military-Learning/Journal-of-Military-Learning-Archives/April-2022/Cates-Action-Review/.

Chi, Michelene T. H., Nicholas De Leeuw, Mei-Hung Chiu, and Christian LaVancher. 1994. "Eliciting Self-Explanations Improves Understanding." *Cognitive Science* 18 (3): 439–477.

Deloitte. 2025a. *Global Process Mining Survey 2025*. London: Deloitte.

Deloitte. 2025b. *Using AI to Fight Insurance Fraud*. New York: Deloitte Insights. https://www.deloitte.com/us/en/insights/industry/financial-services/financial-services-industry-predictions/2025/ai-to-fight-insurance-fraud.html.

Deloitte. 2026a. *State of AI in the Enterprise: The Untapped Edge*. New York: Deloitte. (Survey conducted August–September 2025; published January 2026.)

Deloitte. 2026b. *State of Generative AI in the Enterprise: Q4 2025 Wave*. New York: Deloitte. https://www.deloitte.com/us/en/what-we-do/capabilities/applied-artificial-intelligence/content/state-of-generative-ai-in-enterprise.html.

Deloitte. 2026c. *Unlocking Exponential Value with AI Agent Orchestration*. TMT Predictions 2026. London: Deloitte Insights. https://www.deloitte.com/us/en/insights/industry/technology/technology-media-and-telecom-predictions/2026/ai-agent-orchestration.html.

Dell'Acqua, Fabrizio, Edward McFowland III, Ethan Mollick, Hila Lifshitz-Assaf, Katherine Kellogg, Saran Rajendran, Lisa Krayer, François Candelon, and Karim R. Lakhani. 2025. "Navigating the Jagged Technological Frontier: Field Experimental Evidence of the Effects of AI on Knowledge Worker Productivity and Quality." *Organization Science*. (Working paper version: Harvard Business School Working Paper 24-013, 2023.)

Di Stefano, Giada, Francesca Gino, Gary P. Pisano, and Bradley R. Staats. 2016. "Making Experience Count: The Role of Reflection in Individual Learning." Harvard Business School Working Paper 14-093.

DORA and Google Cloud. 2025. *2025 State of AI-Assisted Software Development Report*. Mountain View, CA: Google Cloud.

https://dora.dev/dora-report-2025/.

Dratsch, Thomas, Xixing Chen, Mohammad Mehrabi Rashidi, Daniel Pinto dos Santos, Andra-Iza Iuga, Birte Kuntze, et al. 2023. "Automation Bias in Mammography: The Impact of Artificial Intelligence BI-RADS Suggestions on Reader Performance." *Radiology* 307 (4). https://pubs.rsna.org/doi/10.1148/radiol.222176.

Edelman. 2025a. *2025 Edelman Trust Barometer*. New York: Edelman. https://www.edelman.com/trust/2025/trust-barometer.

Edelman. 2025b. *Flash Poll: Trust and Artificial Intelligence at a Crossroads*. New York: Edelman. (November 2025.)

Education Endowment Foundation and NFER. 2024. *Choices in Edtech: Using ChatGPT in Lesson Preparation — Teacher Choices Trial*. London: EEF. https://educationendowmentfoundation.org.uk/news/teachers-using-chatgpt-alongside-a-guide-to-support-them-to-use-it-effectively-can-cut-lesson-planning-time-by-over-30-per-cent.

Forrester Consulting. 2025. *The Projected Total Economic Impact of Microsoft Teams with Microsoft 365 Copilot*. Cambridge, MA: Forrester Research. (Commissioned by Microsoft.)

Gallup. 2026. *State of the Global Workplace: Q1 2026 AI Insights*. Washington, DC: Gallup.

Gartner. 2025a. "Gartner Predicts 40% of Enterprise Apps Will Feature Task-Specific AI Agents by 2026, Up from Less Than 5% in 2025." Press release, August 26. https://www.gartner.com/en/newsroom/press-releases/2025-08-26-gartner-predicts-40-percent-of-enterprise-apps-will-feature-task-specific-ai-agents-by-2026-up-from-less-than-5-percent-in-2025.

Gartner. 2025b. "Gartner Predicts Over 40% of Agentic AI Projects Will Be Canceled by End of 2027." Press release, June 25. https://www.gartner.com/en/newsroom/press-releases/2025-06-25-gartner-predicts-over-40-percent-of-agentic-ai-projects-will-be-canceled-by-end-of-2027.

GitHub. 2025. *Octoverse 2025: AI Leads Python to #1*. https://github.blog/news-insights/octoverse/octoverse-2025/.

Hutchins, Edwin. 1995. *Cognition in the Wild*. Cambridge, MA: MIT Press.

IBM Institute for Business Value. 2025. *Cut the Cost of Complexity with Intelligent IT Automation*. Armonk, NY: IBM. https://www.ibm.com/thought-leadership/institute-business-value/en-us/report/intelligent-it-automation.

IBM Security and Ponemon Institute. 2025. *2025 Cost of a Data Breach Report*. Armonk, NY: IBM.

Information Age. 2025. "First Australian Lawyer Penalised for AI Blunder." *Information Age*, Australian Computer Society. https://ia.acs.org.au/article/2025/first-australian-lawyer-penalised-for-ai-blunder.html.

Interactive Advertising Bureau and PwC. 2026. *Internet Advertising Revenue Report: Full Year 2025*. New York: IAB. https://www.iab.com/insights/internet-advertising-revenue-report-full-year-2025/.

James, William. 1890. *The Principles of Psychology*. New York: Henry Holt and Company.

Kwa, Megan, Joshua West, et al. 2025. *Measuring AI Ability to Complete Long Tasks*. Berkeley, CA: METR. (Updated as *Time Horizon 1.1*, January 2026.) https://metr.org/blog/2025-03-19-measuring-ai-ability-to-complete-long-tasks/.

LIONS Advisory. 2025. *The State of Creativity 2025*. London: Cannes Lions International Festival of Creativity. https://www.lions.co/en/state-of-creativity-2025.

Magesh, Varun, Faiz Surani, Matthew Dahl, Mirac Suzgun, Christopher D. Manning, and Daniel E. Ho. 2025. "Hallucination-Free? Assessing the Reliability of Leading AI Legal Research Tools." Stanford RegLab Working Paper. Forthcoming in *Journal of Empirical Legal Studies*. https://reglab.stanford.edu/publications/hallucination-free-assessing-the-reliability-of-leading-ai-legal-research-tools/.

McKinsey & Company. 2025a. *Superagency in the Workplace: Empowering People to Unlock AI's Full Potential at Work*. New York: McKinsey & Company. https://www.mckinsey.com/capabilities/tech-and-ai/our-insights/superagency-in-the-workplace-empowering-people-to-unlock-ais-full-potential-at-work.

McKinsey & Company. 2025b. *The Future of AI for the Insurance Industry*. New York: McKinsey & Company. https://www.mckinsey.com/industries/financial-services/our-insights/the-future-of-ai-in-the-insurance-industry.

McKinsey & Company. 2025c. *The State of AI: How Organizations Are Rewiring to Capture Value*. New York: McKinsey & Company. https://www.mckinsey.com/capabilities/quantumblack/our-insights/the-state-of-ai-how-organizations-are-rewiring-to-capture-value.

McKinsey & Company. 2025d. *The State of AI in 2025: Agents, Innovation, and Transformation*. New York: McKinsey & Company. https://www.mckinsey.com/capabilities/quantumblack/our-insights/the-state-of-ai.

McKinsey Global Institute. 2012. *The Social Economy: Unlocking Value and Productivity Through Social Technologies*. New York: McKinsey & Company. https://www.mckinsey.com/industries/technology-media-and-telecommunications/our-insights/the-social-economy.

MIT NANDA Initiative. 2025. *The GenAI Divide: State of AI in Business 2025*. Cambridge, MA: MIT Media Lab.

MIT Sloan Management Review and Boston Consulting Group. 2025. *Five Insights from Our Ninth Annual Global AI Study*. Cambridge, MA: MIT Sloan Management Review. https://sloanreview.mit.edu/projects/five-insights-from-our-ninth-annual-global-ai-study/.

Nonaka, Ikujiro. 1991. "The Knowledge-Creating Company." *Harvard Business Review* 69 (6): 96–104.

OECD. 2025. *Bridging the AI Skills Gap: Is Training Keeping Up?* OECD

Artificial Intelligence Papers. Paris: OECD Publishing.

Pew Research Center. 2025. *U.S. Workers Are More Worried Than Hopeful About Future AI Use in the Workplace*. Washington, DC: Pew Research Center. https://www.pewresearch.org/social-trends/2025/02/25/u-s-workers-are-more-worried-than-hopeful-about-future-ai-use-in-the-workplace/.

Project Management Institute. 2018. *Pulse of the Profession 2018: Success in Disruptive Times*. Newtown Square, PA: PMI.

Prosci. 2024. *Best Practices in Change Management*, 11th ed. Fort Collins, CO: Prosci.

Rogelberg, Steven G. 2022. "Dear Manager, You're Holding Too Many Meetings." *Harvard Business Review*, March 8. https://hbr.org/2022/03/dear-manager-youre-holding-too-many-meetings.

Ruan, Sherry, Jacob O. Wobbrock, Kenny Liou, Andrew Ng, and James A. Landay. 2017. *Comparing Speech and Keyboard Text Entry for Short Messages in Two Languages on Touchscreen Phones*. Stanford HCI Working Paper.

Schubel, Lauren C., et al. 2025. "Ambient AI Scribes in Clinical Practice: A Randomized Trial." *NEJM AI* 2 (3). https://ai.nejm.org/doi/full/10.1056/AIoa2501000.

Stack Overflow. 2025. *2025 Developer Survey: AI Section*. New York: Stack Overflow. https://survey.stackoverflow.co/2025/ai.

Stanford Institute for Human-Centered Artificial Intelligence. 2025. *2025 AI Index Report*. Stanford, CA: Stanford University. https://hai.stanford.edu/ai-index/2025-ai-index-report.

Thomson Reuters Institute. 2025. *Future of Professionals Report 2025*. Eagan, MN: Thomson Reuters.

US Food and Drug Administration. 2025. *Artificial Intelligence-Enabled Medical Devices*. Silver Spring, MD: FDA. (List current to 30 December 2025.) https://www.fda.gov/medical-devices/software-medical-device-samd/artificial-intelligence-enabled-medical-devices.

USC Annenberg Center for Public Relations. 2025. *2025 Global Communication Report: Mind the Gap*. Los Angeles: University of Southern California. https://annenberg.usc.edu/research/center-public-relations/global-communication-report.

Wang, Rose E., Dorottya Demszky, Matthew A. Kraft, et al. 2024. *Tutor CoPilot: A Human-AI Approach for Scaling Real-Time Expertise*. Stanford SCALE Initiative / National Student Support Accelerator. https://nssa.stanford.edu/studies/tutor-copilot-human-ai-approach-scaling-real-time-expertise.

World Economic Forum. 2025. *Future of Jobs Report 2025*. Geneva: World Economic Forum. https://www.weforum.org/publications/the-future-of-jobs-report-2025/.

www.ingramcontent.com/pod-product-compliance
Lightning Source LLC
La Vergne TN
LVHW052034200926
848481LV00005B/1679